Paul Murdin, OBE, is the author of *The Secret Lives of Planets*. He has worked as an astronomer in the UK, the USA, Australia and Spain, and discovered the first stellar black hole, Cygnus X1. He is a fellow of the Royal Astronomical Society, former president of the European Astronomical Society and Senior Professor Emeritus at the Institute of Astronomy at the University of Cambridge. In 2012 Asteroid 128562 was named 'Murdin' by the IAU in honour of his contributions to the field.

D1741343

PAUL MURDIN

SECRETS
OF THE
UNIVERSE

How We Discovered the Cosmos

Revised edition

With 29 illustrations

On the cover: Picture of Neptune taken by
the Hubble Space Telescope. Astronomers used
different colour filters to detect atmospheric
features that cannot be seen when looking at
the natural colours. Photo: Shutterstock.

First published in the United Kingdom in 2009
by Thames & Hudson Ltd, 181A High Holborn,
London WC1V 7QX

This revised edition 2020

Secrets of the Universe: How We Discovered the Cosmos
© 2009 and 2020 Thames & Hudson Ltd, London

Text by Paul Murdin

British Library Cataloguing-in-Publication Data
A catalogue record for this book is available from
the British Library

ISBN 978-0-500-29519-9

Printed and bound in the UK by CPI (UK) Ltd

To find out about all our publications, please visit
www.thamesandhudson.com. There you can subscribe
to our e-newsletter, browse or download our current
catalogue, and buy any titles that are in print.

CONTENTS

Introduction

The Australian Astronomical Observatory is located on a mountain near the small town of Coonabarabran in rural New South Wales, Siding Spring, in the Warrumbungle range, on the edge of a national park 400 kilometres northwest of Sydney. The seemingly interminable plains of the outback stretch to the western horizon. In the foreground, across a deep cut valley, are the volcanic hills, dykes and plugs of the range, with fanciful names: the Breadknife, Belougery Spire, Crater Bluff. Eucalyptus trees stand tall on the steep sides of the mountains, difficult for all to clamber through, except for kangaroos, koalas and wallabies. Brightly coloured lorikeets, cockatoos, galahs and rosellas fly singly and in flocks through the tree canopy, quarrelling with each other and their neighbours. The telescope buildings contrast with this natural scene; their hemispherical domes are painted white to reflect the heat of the Sun during the day in order to keep the telescopes cool at night. The whole area lies within a light-pollution protection zone, and the skies above the telescopes are in one sense dark, in another sense brilliant with the splendour of the Southern Hemisphere stars, especially in the southern winter when the centre of the Milky Way arches across the zenith.

Right after the telescope was completed, I was lucky enough to join the first group of six scientific staff of the Observatory, then called the Anglo-Australian Observatory. Its 4-metre telescope was at that time the largest in the Southern Hemisphere. It was built to a very high specification, and equipped with sensitive instrumentation, including what were then innovative computer-controlled electronic detectors. What it revealed, almost wherever it pointed, were new discoveries.

I well remember one such discovery. I had been working for twelve hours non-stop through the night on the identification of an X-ray source – that is, I set myself with the telescope to track down the optical star that was responsible for a beam of celestial X-rays that had been detected by a satellite with a telescope sensitive to X-rays.

I found it. I had also discovered, fairly clearly, that the star was the result of a relatively recent supernova explosion that had taken place perhaps 3,000 years ago. I was also able to estimate that the star was about 2,000 light years away.

I finished the night's work as the day dawned, helped the telescope operator to shut down and left the observatory building in the golden light of the rising Sun to walk to the Lodge and a welcome sleep. The kangaroos and wallabies were finishing their night's grazing and loped away unhurriedly from my path across the grass, towards their own beds in the bush. Kookaburras were greeting the dawn with mad laughter, which echoed in the valleys below. Large, black birds called currawongs were waking up in the gum trees with a dawn chorus of melodious warbling and chortling.

I was feeling tired, but it was a lovely morning and I had had a successful night. I was the only person in the world who knew what I knew about the star and I felt pretty good.

As I walked along the path an even more thrilling thought struck me. Light from the explosion of the supernova had taken 2,000 years to get to Earth, and it had travelled on for a further 3,000 light years. It was now ranged on the surface of a sphere 5,000 light years in radius. Outside this sphere the supernova had not yet happened.

Now, 5,000 light years sounds like a large distance – and indeed it is. However, it is not large compared with the size of the Galaxy, and although there are some stars within this sphere, there are not many. Some of them, like our Sun, may have planets, and some of the planets may be habitable, and some of them may have life on them, and some of this life might be intelligent, and some of the civilizations on these planets might be interested in astronomy and might have seen the supernova – or, as I had, the X-ray source – and followed the trail to the interesting star. But there was also a good chance that there was only one such civilization – the one here on Earth. If there was indeed only the one astronomy-curious civilization within the 5,000-light-year sphere – our own – I was not only the one person in the world who knew what I knew, I might be the one person in the Galaxy, or indeed in the Universe, who knew what I knew. I floated happily to bed and, pleased with myself, slept soundly, treasuring in my dreams the secret that I had unlocked from the Universe.

I was happy to learn during the research for this book that other scientists had felt the same sense of exhilaration when they had cracked a problem and made a cosmic discovery. Einstein's worry about his own theory of General Relativity was assuaged by his discovery of the reason for a twist in the orbit of Mercury and, for a few days, he was beside himself with 'joyous excitement'. Henry Norris Russell described the discovery of white dwarfs with Edward Pickering and Williamina Fleming: 'I knew enough, even then, to know what it meant... At that moment, Pickering, Mrs Fleming and I were the only people in the world who knew of white dwarfs.' Watching the transit of Venus in 1639, William Crabtree 'stood for some time motionless, scarcely trusting his own senses, through excess of joy'.

Discoveries in astronomy challenge our fundamental assumptions about the Universe. They alter our perception of matter, time and distance; they transform how we view our history and future as a species. Where the astronomers of antiquity spoke of fixed stars, we speak of whirling galaxies and the death and birth of stars in supernovae. Where we once considered the Earth to be the centre of the

Universe, we now see it as a small planet among millions of similar systems, a few of which might also hold life. These dramatic shifts in perspective hinge on thousands of individual moments of discovery, moments when it became clear to an observer that a component of the Universe – from a tiny subatomic particle to a supermassive black hole – was not as it once seemed. Each is a revelation that unlocks yet another of the infinite secrets of the Universe.

My own discoveries were more modest, at the scale of the quanta of astronomy – over 50,000 pages per year of similar-scale discoveries in astronomy are published every year. This book is about major discoveries that unlocked major secrets of the Universe – not run-of-the-mill discoveries in astronomy, but the big ones. I have selected them not only because they are important, but also because the people involved were interesting, because the knowledge that they encompass illustrates the range of astronomy, or because the stories behind them illustrate how science, and in particular astronomy, works.

Science is a cyclic process that oscillates between the real world ('observation' or 'experiment') and the picture of it in the scientist's head ('theory'). A scientific discovery might be the revelation of one of the secrets of the Universe by finding something in the real world. A secret might also be revealed by bringing together such a compelling set of interpretations that the picture in one person's head is accepted by most people as a portrait of the real world. Galileo saw mountains on the Moon. Copernicus pictured the Sun to be at the centre of the Solar System. Both were discoveries; one of them an observation, one of them a theory.

For the man in the street, the word 'theoretical' sometimes carries a sense of derision – you can't trust something that is 'just a theory'. For scientists, the word can mean something that is as solid as the chair on which I am sitting, so theories can definitely be discoveries too. But perhaps to be a discovery a theory needs an extra something. Either it brings together a number of previously unrelated phenomena with such clarity that everybody is convinced it is right, or it points to some phenomenon that has not been seen yet but which turns out, when looked for, to be so.

The word 'discovery' carries the implication that what is now known was not known before, perhaps not even suspected. It is a bit of a surprise. The most straightforward reason in astronomy for such a surprise discovery is that, like the discoveries that I made with the AAT, it was made possible by some improvement in technical capability. Galileo learned how to make a telescope and used it to point to the sky. What he discovered – the satellites of Jupiter, the phases of Venus, star clusters – confirmed not only that the Sun, not the Earth, was the centre of the Solar System, but also that everything in space operated in much the same way that things did on Earth; there is no difference in principle between the 'mundane' (terrestrial things) and the 'superlunary' (things beyond the Moon and therefore cosmic). We on Earth are not apart from the Universe, but a part of the Universe.

William Herschel built bigger telescopes that opened the window wider and wider on the Universe; the Hubble Space Telescope blasted open the doors. The development of radio astronomy, X-ray astronomy and infrared and millimetre-wave astronomy in the twentieth century allowed us to see objects in the Universe that are invisible to the naked eye. Martin Ryle's invention of the technique of Aperture Synthesis Interferometry in radio astronomy made it possible to investigate radio galaxies, which showed that the Universe had a discrete beginning. When they reached the right level of sensitivity, gravitational-wave detectors opened up a completely new window on the Universe. Since 1957, the possibility for spacecraft to carry equipment to the distant reaches of the Solar System has offered new perspectives on the planets.

Discoveries with new equipment are in one sense unexpected but in another sense planned, because the equipment has to be made and deployed. That means having the right idea, gathering together the resources, and carrying out a plan to use the equipment for a specific purpose. William Herschel built a new telescope and found the planet Uranus by systematically searching the sky with it; his sister, Caroline, found her comet by applying the same technique. In modern times the new equipment has to be bought, and that takes money – sometimes lots – so a detailed funding application has to be written, predicting what discoveries will be made with this expensive new telescope or satellite. If you just tell the truth – that the Universe is full of exciting things, and you can find something interesting with every instrumental advance – well, you won't get funding. You have to at least scope the range of potential discoveries to show that you are serious about your research.

Certainly, in some cases scientists set out to find something specific. Urbain Le Verrier perceived the planet Neptune 'at the end of his pen; he determined it by the mere force of calculation', while Daniel Barringer became obsessed with the idea that the Coon Butte crater in Arizona was meteoritic, spurred on by the thought of finding a profitable mass of iron and nickel. Subrahmanyan Chandrasekhar calculated the structure of white dwarf stars as a student exercise that he set himself to pass the time on an ocean voyage and uncovered the reason for black holes. Raymond Davis spent over ten years searching for neutrinos from inside the Sun; his discovery led to the development of a new kind of physics, properly deserving of the award of the Nobel Prize.

Computer modelling has shed new light on known phenomena, making surprising new astronomical discoveries possible. The phrase 'Garbage in, garbage out' is well known; you might think that its corollary is 'Put in the truth that you know, get out the truth that you know.' But when there is a lot of data or the calculations are complex, computers can reveal unexpected or previously unnoticed features about the real Universe. Computer simulations of the interactions of asteroids and comets led to our present understanding of the Oort Cloud and the Kuiper Belt. Satellites have great difficulty in probing the magnetosphere of the Earth because it is so large that they can only investigate particular parts of it, like the people in the fable who grope the tail, foot, tusks, and trunk of an elephant, but fail to envisage the whole creature; computers are able to assemble these fragments into a complete picture. The Universe is hard to study because you can't compare and contrast it with other real universes, but the Millennium Simulation models universes that are different from ours, which helps us to estimate how much dark matter and dark energy there are in the real Universe.

Sometimes astronomical discoveries are serendipitous: the right person is in the right place at the right time. Tycho Brahe was returning home from an evening dinner at the time that the supernova of 1572 appeared in the sky and saw it while gazing from his carriage window; 400 years later, Ian Shelton happened to be pointing his telescope in the right direction when Supernova 1987A exploded. The crucial factor was that both discoverers knew about astronomy and understood what they were seeing. Other cosmic discoveries were unexpected by-products of investigations set up for entirely different purposes. Herb Gursky and Riccardo Giacconi made a serendipitous discovery when an X-ray detector on a rocket they had launched to look at the Moon saw a bright source behind it. Jocelyn Bell discovered pulsars as a source of 'noise' ('scruff', as she called it) during the observation of quasars. In both cases the scientists were remarkably persistent in systematically tracking down the origin of the inconsistency.

In a lecture in Lille in 1854 the French scientist Louis Pasteur perceptively noted that 'in the field of observation, chance favours only the prepared mind.' Usually in astronomy this means the prepared multidisciplinary mind. Astronomy as a subject encompasses the study of everything in space. Physics, mathematics, chemistry, computing, engineering, statistics – all these sciences, and more, are deployed by astronomers to understand what they see and to unlock cosmic secrets. Some of the most important discoveries are collective, the product of investigations made by many different people over several generations, although there is usually one last genius who ties it all together. The laws of the motion of the planets engaged the minds of a stellar system of talents before the proverbial apple dropped and Isaac Newton discovered the theory of gravity. 'If I have seen further, it is by standing on the shoulders of giants,' he wrote. The discovery of the greenhouse effect in the atmospheres of Venus and the Earth took 150 years of investigation by dozens of scientists – there was really no one person who made the discovery, now recognized as so momentous that nothing less than the survival of life on Earth may depend upon our understanding of it. By contrast, Special Relativity and General Relativity were the ideas of a single individual, Albert Einstein, working over only a few years.

'The most exciting phrase to hear in science, the one that heralds new discoveries, is not "Eureka!", but "That's funny...".' This saying, commonly attributed to science-fiction writer Isaac Asimov, captures the sense that the most important feature of a scientific discovery is the open mind and curiosity of the person who makes it. In this book I have tried to explain what lies behind some of the great discoveries in astronomy, the train of events and thoughts that brought the scientist to exclaim 'Eureka' or 'That's funny...' as he or she unlocked one of the major secrets of the Universe. This book is therefore mostly scientific history – my top cosmic discoveries. However, in the first edition of the book I also identified four major prospects for future discoveries: to uncover the secrets of dark matter and dark energy, to detect gravitational waves and to discover life elsewhere in the Universe – although we may not find what we expect. The challenge for the next generation of astronomers, I wrote in 2008, was to put themselves in the position to uncover these momentous secrets. I hoped that some of them might succeed in my lifetime.

I am lucky enough to have seen one of these four predicted discoveries come to pass: the discovery of gravitational waves. Gravitational waves were first identified in 2015, originating from two colliding black holes. Their existence had been expected – but, as usual, the Universe is cleverer than astronomers, and there was an entirely unexpected feature of the discovery. I have no doubt whatsoever that there will be surprises in the other three future discoveries that I have anticipated, not to mention all those discoveries that I have not written about but are surely still to come.

Paul Murdin, 2019

BEFORE THE

TELESCOPE

The Seven Planets
Wandering stars

Day 13 [20 September]. Sunset to moonrise: 8°. There was a
lunar eclipse. Its totality was covered at the moment when
Jupiter set and Saturn rose. During totality the west wind
blew, during clearing the east wind. During the eclipse, deaths
and plague occurred. That month, the equivalent for 1 *shekel*
of silver was: barley, [so many] *kur*; mustard, 3 *kur*; sesame,
1 *pân*, 5 *minas*. At that time, Jupiter was in Scorpio; Venus
was in Leo, at the end of the month in Virgo; Saturn was in
Pisces; Mercury and Mars, which had set, were not visible.

Cuneiform tablet, 331 BCE

The human fascination with the movement of the planets is almost
as old as humanity itself. 25,000 years ago, beside a lake in Africa, a
member of the Ishango community recorded the cycles of the Moon;
in ancient Babylon, astronomers used their knowledge of the planets
to advise their king about affairs of state and the price of barley. For
2,000 years the 'geocentric theory' provided the dominant explanation
of the Universe, based on the evidence of everyday human observations.

The word 'planet' comes from the Greek word for 'wanderer',
because originally any celestial objects that were not 'fixed' stars
were considered to be planets. The Sun and Moon were thought to
be planets and their motions could, with careful attention and the
keeping of records, be predicted. It seemed reasonable to assume
that the same was true of Mercury, Venus, Mars, Jupiter and Saturn,
which were visible to the naked eye as bright lights roaming across
the heavens. To many ancient observers of the skies it seemed that,
if they could understand and predict the motions of the seven
planets, they might uncover the deepest mysteries of the cosmos.

We have a tangible connection with one of the first known observers of the Moon, 25,000 years ago. He or she was one of a community known as the Ishango people who, until they were dispersed or wiped out by a volcanic eruption, lived, fished and farmed along the shores of what we now call Lake Edward, one of the sources of the Nile in central equatorial Africa. This person scratched markings in groups of twenty-nine on the bone handle of a knife or chisel. These markings have changing sizes that, according to anthropologist Alexander Marshack, represent the varying phases of the Moon throughout six months of the lunar cycle. There are gaps in the markings that seem to represent cloudy nights when the Moon was not seen.

We know nothing more about the maker of the Ishango bone – the earliest surviving lunar calendar. Perhaps the carver was a hunter or traveller keeping track of a long journey, perhaps a woman keeping track of her menstrual cycle.

Across the continents the erection of monuments and, where they have survived, the evidence of ancient myths and legends reveal the many different ways in which prehistoric men and women made sense of the cosmos and its relationship to their lives. It is unlikely that these peoples made precise astronomical observations, even at an elaborate primitive observatory like Stonehenge, but the keeping of calendars that tracked the solar and lunar cycles required the first civilizations to develop more complex patterns of thought and communication. Modern physicists use advanced mathematics to explain the operation of the Universe. It was the observation of the heavens that encouraged humankind to take the earliest steps towards mathematical thought.

The first known systematic observations of the planets were recorded on Babylonian cuneiform tablets, starting from about 1700 BCE. In the Neo-Assyrian period (911–612 BCE) astronomers regularly recorded the motions of the planets and, based on what they saw, gave astrological forecasts to the king. Astronomers of the Achaemenid kingdom kept astronomical diaries of their observations, which they used to make predictions about affairs of state, the level of the River Euphrates, and the price of goods such as barley, dates, mustard, sesame and wool.

Alexander the Great's conquests in the East in the fourth century BCE brought the Babylonians' detailed astronomical records to the attention of ancient Greece, where philosophers such as Thales of Miletus, Pythagoras, Plato and Aristotle had debated the nature of the Universe and used geometry to explain planetary motions. The Babylonians' example encouraged Greek astronomers to base their speculations upon more exact observations of the stars and planets.

The Greeks assumed that the Earth, which was not a planet, lay stationary at the centre of the Universe – after all, the Earth doesn't rock about as if it is moving, nor do the positions of the stars change as if we are viewing them from a succession of positions along an orbit. This was the 'geocentric' theory of the planets and it is associated with the name of the astronomer Claudius Ptolemaeus, more often known as Ptolemy.

Ptolemy worked at or near Alexandria in Egypt during the middle decades of the second century CE. By 147 CE he had developed the geocentric theory to a sophisticated level, described in outline in a public inscription and presented in full in a large treatise entitled the *Almagest*, written in Greek in the second century CE, but now known only through Arabic versions that were translated into Latin in the Middle Ages (its title, meaning 'the greatest', comes from an Arabic translation).

The Ptolemaic theory of the Universe held that the Moon, Mercury, Venus, the Sun, Mars, Jupiter and Saturn revolved in a succession of concentric orbits. They orbited around the Earth as it lay motionless at the Universe's centre. Although astronomers had by this time introduced the idea of orbits, the picture that was commonly in mind was that the planets were mounted on a series of hollow crystal spheres. Outside the sphere of the most distant planet, Saturn, was the sphere of the stars – because at any time of the night the stars seem to be studded on a hemisphere above us and this sphere appears to rotate around the Earth. (Navigators still use the convention of the celestial sphere to calculate the positions of the stars as seen from ships.) Beyond the sphere of the stars was an unseen sphere – the 'primum mobile' ('first mover') – which was the primary mechanism that drove the movements of the Universe.

What were the engines that rotated the celestial spheres on their axes? The Christian astronomers who adopted Ptolemy's theories eventually offered their own answer: eight of Gabriel's angels pushed the spheres through their rotations.

In its essentials, the geocentric theory of the Universe is the common-sense view of its structure: it is what we feel and see. The true arrangement of the Universe would remain hidden from human perception for nearly 2,000 years.

Stars and Northern Constellations
Our human link with the Ice Age

> The whole of mythology could be taken as a sort of projection
> of the collective unconscious. We can see this most clearly
> if we look at the heavenly constellations, whose originally
> chaotic forms are organized through the projection of
> images. This explains the influence of the stars as asserted
> by astrologers. These influences are nothing but unconscious,
> introspective perceptions of the collective unconscious.
>
> Carl Jung, *The Structure of the Psyche*, 1927

Although the stars are random in position and brightness, they form shapes that people can interpret symbolically, just as they read clouds, tea leaves or entrails. Even before recorded history, people have seen mythical heroes, animals, birds and everyday objects in the patterns of the stars. The names of these ancient constellations are still used by astronomers, and are among the oldest surviving elements of human culture.

One of the most ancient constellations is Ursa Major, which was identified as a bear by northern peoples in both North America and Eurasia, and must therefore have pre-dated the disappearance of the land-bridge joining Alaska and Siberia around 15,000 years ago. The concept of the constellation as a bear then spread southward into the Middle East and the eastern Mediterranean, even though bears had disappeared from these lands as the glaciers retreated at the end of the last Ice Age. The Pointers of Ursa Major help to locate the North (pole) Star in the sky, and for this reason the Great Bear is often the first constellation that people in the Northern Hemisphere learn. Ursa Major symbolizes the northern regions and features on the flags of Alaska and the Cherokee nation.

Forty-five constellations were described in the *Phaenomena*, a poem written in about 275 BCE by the Macedonian Greek Aratus of Soli. The poem was itself based on earlier work, now lost, by Eudoxus, a Greek astronomer and mathematician of the fourth century BCE. Eudoxus pored over old manuscripts in libraries in Egypt, which recorded original constellations dating back to the Babylonian civilizations of Mesopotamia. These manuscripts, too, are lost (some in the fires that destroyed the Library of Alexandria by the fifth century CE), but the constellations they recorded survived in Greek literary culture. By the age of the poet Homer, the Babylonian constellations had become interwoven with Greek mythology, a process that was virtually complete by the third century BCE.

Most of the ancient constellations acquired their present Latin names in the first and second centuries as Greek mythology became absorbed into Roman culture and Latin translations of Greek texts like *Phaenomena* appeared. Representations of the constellation figures appeared: the oldest surviving representation is the statue known as the Farnese Atlas. Sentenced by Zeus to hold up the celestial sphere, the Titan Atlas struggles under its weight in this second-century Roman marble copy of a Hellenistic sculpture. The globe shows forty-one of the classical Greek constellations listed by Aratus and is the oldest surviving picture of them. There has been speculation that the constellations in this depiction are based on the lost star catalogue by Hipparchus dating from 129 BCE. The Greek-speaking astronomer Ptolemy, who lived in the Roman colony of Alexandria in Egypt, described forty-eight constellations in his *Almagest*. These descriptions of the constellations formed the basis for the constellations of the present day.

The oldest surviving map of the stars on paper is Chinese, and is known as the *Dunhuang Star Chart*, dating from about 720 CE. It identified four hundred Chinese asterisms (sub-constellations): they are smaller and more numerous than Western constellations, and not usually related to them, although some are recognizable, such as the Seven Stars of the Northern Dipper (in Chinese, Beidou

Qixing), part of Ursa Major. This star chart is one of tens of thousands of scroll books that were discovered in 1908 in a library in the Mogao Buddhist caves near Dunhuang in northwest China, hidden in a room that had been bricked up for a thousand years in anticipation of the imminent arrival of marauders.

In the West, a number of constellations have been added into Ptolemy's list, filling both the gaps between the Greco-Roman constellations and the large gap in the southern skies that was not visible from Mediterranean latitudes. The Polish astronomer Johannes Hevelius named seven constellations in 1687. They included Lacerta, since, as Hevelius wryly reasoned, only a lizard could wriggle into the small space available, and the Lynx, because the sharp eyes of a lynx were needed to see its faint stars. Some astronomers attempted to honour patrons by naming constellations for them, but only one survives: Hevelius's constellation Scutum was originally Scutum Sobiescianum, named after Poland's King John III Sobieski.

Other modern constellations never found widespread favour and were included on some charts but ignored on others. In 1922 the International Astronomical Union (IAU) took charge of the chaotic situation. Under the Belgian astronomer Eugène Delporte, the IAU standardized the constellations to the official modern system, abbreviating some over-elaborate names, rendering many constellations obsolete. There are now eighty-eight recognized constellations with fixed boundaries.

Alongside the official names of the constellations, there are common names in use in various languages. Gemini, for example, is known as the Twins in English, Gémeaux in French, Zwillinge in German and Gemelli in Italian. There are also names for asterisms, such as the Plough, the Big Dipper and Charles' Wain (all actually naming the same part of Ursa Major) and the Hyades (a star cluster in the constellation Taurus). The Summer Triangle of the northern sky (an equilateral triangle formed by the stars Altair, Deneb and Vega) is an asterism that spans several constellations. Some Australian Aboriginal peoples saw a constellation in the

Southern Cross that was not made of stars, but of dark clouds in the shape of an emu.

The zodiacal constellations are the major constellations that the Sun, Moon and planets cross as they travel around the sky. These constellations are of different sizes, but give their names to a system of twelve abstract 'zodiacal signs' that divide the sky evenly into sectors that are 30 degrees long and which provide a way of describing the positions of the Sun and planets. When the system was first devised, about 3,000 years ago, there were six zodiacal constellations, all of them representing animals, such as Leo, the Lion – hence the name of 'zodiac', a Greek word meaning 'figured like animals'. Six further constellations were later interpolated, some of them inanimate, such as Libra, the Scales. The twelve zodiacal signs were named after the twelve zodiacal constellations. This location system for planets constitutes the framework used by astrology, which bases its predictions on planetary positions. The zodiacal signs emerged in Babylonian astronomy during the fifth century BCE and travelled to Greece, Egypt, Rome and India. Because the Solar System tilts, due to the phenomenon of 'precession', a thirteenth constellation, Ophiuchus, now intrudes into the zodiac, but this has not affected the system of zodiacal signs, and astrology remains largely indifferent to it. The continuing popularity of consulting one's 'star signs' in daily horoscopes attests to the lasting influence of the ancient constellations on the public imagination, even though the prophetic powers of astrology are pseudo-scientific nonsense.

The Milky Way
Path of the gods, souls and pilgrims

> He had gauged the southern skies with greater results
> than even he himself had anticipated. Those unfamiliar
> constellations which, to the casual beholder, are at most a
> new arrangement of ordinary points of light, were to this
> professed astronomer, as to his brethren, a far greater
> matter. It was below the surface that his material lay. There,
> in regions revealed only to the instrumental observer, were
> suns of hybrid kind – fire-fogs, floating nuclei, globes that
> flew in groups like swarms of bees, and other extraordinary
> sights – which…turned out to be the beginning of a new
> series of phenomena instead of the end of an old one.

Thomas Hardy, *Two on a Tower*, 1882

Our Sun is part of a disc-shaped collection of stars called the Milky Way Galaxy, or just 'the Galaxy'. The Sun lies in the plane of the disc. When we look at the sky along the direction of this plane, we see large numbers of stars, massed into a filmy band. This is the Milky Way, which has the appearance of a milk-like stream of light, although it is not actually a band or a stream, but a disc of stars seen edge-on.

The earliest surviving written description of the Milky Way is in Ptolemy's *Almagest*. 'The Milky Way is not simply a circle,' he wrote, 'but a zone having almost the colour of milk, whence its name. It is not regular and ordered but different in width, colour, density and position; in one part it is double.' Of course, the very word 'milky' suggested the literal mythological explanation, a favourite scene of painters, Tintoretto among others, that the Milky Way was the milk of the goddess Juno (plate I). The fourth-century BCE Greek scientist and philosopher Aristotle made the first scientific discussion about the Milky Way in his *Meteorologica*, written around 350 BCE, and many Western ideas about the origin or interpretation of the Milky Way can be traced back to this source, in which he discussed and classified all the theories of his time.

Because we live centrally in the plane of the disc of our galaxy, the direction of greatest number of stars in the Milky Way stretches in a great circle around the sky. The Milky Way thus has the appearance of an arch, bridge or road across the night sky, conspicuous from dark locations both in summer and in winter evenings. Near the polar regions, the summer nights are short and twilit so Swedes, for example, can scarcely see the Milky Way during the summer and view it only in the long nights of the winter season. Swedes call the Milky Way Vintergatan – 'Winter Street'.

The Milky Way is bifurcated in some of its sections, giving it the appearance of a meandering river split by dark, elongated islets. This is caused by a thin, crinkled layer of dust, which from Earth is seen edge-on, silhouetted against the Milky Way and hiding an irregular zone of light originating from the broader disc of the stars beyond. Because of this visual effect, in Arab lands the Milky Way was known as Al Nahr ('the River'), but the name was never adopted into Western languages because of the potential for confusion with a long meandering constellation with the name of a river, Eridanus.

The Milky Way has been named, too, after specific roads. English names for the Milky Way include Watling Street, after the Roman road built from Chester to London and on to Dover, and Walsingham Way, after the road from London to the Virgin's shrine in Norfolk. Local people thought that the Milky Way pointed to the shrine. In Spain the Milky Way is sometimes called El Camino de Santiago. This is the pilgrim's track to Santiago de Compostela in northern Spain. Both pilgrims' ways were thronged with travellers, as the Milky Way is with stars.

In many cultures the Milky Way symbolized the journey of the soul into the afterlife. Coupling the belief that the stars represent souls with the idea that the Milky Way was a road, the Romans described the Milky Way as the path to the Seats of the Heroes, 'Heroum Sedes', walked by the departing souls of illustrious men. In *Timaeus* (*c.* 360 BCE), Plato described how

> after having thus framed the universe, [the supreme divinity] allotted to it souls equal in number to the stars, inserting each in each...and he declared also, that after living well for the time appointed to him, each one should once more return to the habitation of his associate star, and spend a blessed and suitable existence.

The similarity of the pale appearance of the Milky Way to grey ash produced another class of explanations: that it was the scorched path of the Sun or, in classical mythology, the disastrous route travelled by Phaeton, son of the sun-god Helios, when he lost control of his father's chariot, which carried the Sun across the sky. Jupiter put a stop to this potential disaster by throwing a thunderbolt, causing Phaeton to hurtle in flames into the river Eridanus. The final theory mentioned by Aristotle was that the Milky Way was some sort of manufacturing imperfection, like the seam of a metal casting or the sewed seam around a leather-covered ball.

The true explanation for the Milky Way's 'milk-like' appearance is that it is made up of many stars, which are too faint and too close to each other to be viewed individually. This was first conjectured in the fifth century BCE by the Greek philosopher Democritus and finally proved by Galileo with his telescope in 1610.

The Shape of the Earth
Our planet, a flattened sphere

> A Frenchman who arrives in London will find
> philosophy, like everything else, very much changed
> there....In Paris, you think of the Earth as a melon;
> in London it is flattened on both sides.

Voltaire, *Letters from England*, 1731

The ancient discovery that the Earth was round rather than flat was only the beginning. Over the past 2,000 years, a clock pendulum, a librarian's horseback journey to Alexandria and a wayward satellite have helped us to establish the exact size and shape of the Earth. It turns out that our planet is not a perfect sphere, but resembles a squashed, dimpled golf ball.

Since antiquity, every educated person has known that the Earth is approximately spherical. The question of whether Christopher Columbus would fall off the edge of the world if he sailed westwards from Spain was founded in ignorance. The real doubts centred on whether he would survive the dangers of the journey (unpredictable weather, sailing hazards, sea monsters) and whether he would be able to discover an alternative route to the East Indies. As history tells us, Columbus survived the ocean crossing and in 1492 landed in what he called the West Indies, actually part of the Americas.

Ancient astronomers argued that the Earth was spherical because the shadow that the Earth cast on the Moon during a lunar eclipse was always circular. It was well known, too, how a lookout at the top of a ship's mast would see land before his fellow sailors on deck, because the lookout was able to see over the curvature of the Earth. Greek philosophers of the Pythagorean school in the sixth century BCE disseminated these standard arguments for a spherical Earth throughout the educated world.

In the third century BCE, Eratosthenes, the librarian of Alexandria, determined the size of the Earth. He had heard that at Syene (present-day Aswan) in Upper Egypt the Sun was directly overhead at noon on the day of the summer solstice – the Sun's rays reached the bottom of a deep well. He determined the length of the shadow of a vertical post at Alexandria on the same day and found that the angle of the Sun was $\frac{1}{50}$ of a circle to the south of the zenith. It is said that he determined the distance between Syene and Alexandria by driving a carriage between the two cities and counting the revolutions of the wheels. He multiplied this distance – 5,000 *stadia* – by 50 to calculate that the circumference of the Earth was 250,000 *stadia*. The modern equivalent of a *stadium* is not securely established, but Eratosthenes' figure is thought to be the equivalent of about 45,000 kilometres, remarkably close to the modern measurement of the Earth's circumference as 40,000 kilometres.

However, by the seventeenth century CE it had become clear that the Earth was not perfectly spherical. The first evidence came from a clock pendulum. In 1671 the Paris Academy of Sciences sent Jean Richer to Cayenne in Guyana, South America, on the Equator, to observe the close approach of Mars to the Earth in 1672 in order to establish its distance and thus the scale of the Solar System. To accomplish this, Richer needed an accurate clock. He took a clock with a pendulum that had beaten seconds correctly in Paris. However, in Cayenne the same clock ran slow and lost two and a half minutes every day. To make it beat seconds accurately in Cayenne, Richer discovered that he had to shorten the pendulum by about 3 millimetres. The reason for this was a mystery.

In 1687 Isaac Newton offered a solution to Richer's discovery that a pendulum beat slower at the Equator than in France: the Earth is not exactly spherical, but bulges at the Equator and is squashed at the poles. It rotates once every twenty-four hours, and the resulting centrifugal force raises the region along the Equator. For the same reasons, gravity is reduced at the Equator, which is why Richer's pendulum beat more slowly in Cayenne.

During the eighteenth century Newton's explanation was confirmed by a massive body of work, chiefly organized by the Academy of Sciences of Paris, and initiated partly as a test to decide between Newton's theory of gravity (which suggested that the Earth was flattened at the poles, like a tangerine) and Descartes' theory of gravity (which could be read to suggest that it was pointy at the poles, like an American football or rugby ball). The measurement was also intended to provide a universal standard of length, the metre, defined in terms of the Earth's circumference. The magnitude of the task is represented by the long list of astronomers involved: the four astronomers of the Cassini dynasty (Jean-Dominique Cassini, his son Jacques Cassini, his grandson César-François Cassini de Thury and his great-grandson, also called Jean-Dominique Cassini, all four of them being successive directors of the Paris Observatory); the astronomers Jean-Baptiste-Joseph Delambre and Pierre Méchain; the French mathematicians Pierre Maupertuis, Pierre Bouguer and Louis Godin; the explorer Charles-Marie de La Condamine; and the Swedish astronomer Anders Celsius, whose name is remembered in the centigrade temperature scale. These astronomers carefully measured the three-dimensional shape of France. Some of them participated in adventures to the Equator in Ecuador and to the Arctic Circle in Lapland to measure the curvature of the Earth. They discovered that the Earth was indeed a flattened sphere. The modern value for the flattening is $1/298.25$ – there is a difference of more than 21 kilometres between the Earth's radius at the equator (6,378 kilometres) and its radius at the pole (6,357 kilometres).

The shape of the Earth affects the motion of orbiting satellites – lumps that spoil the uniform spherical shape pull satellites off course. Some satellites, like the LAGEOS, or Laser Geodynamics Satellites, are simple spheres covered with reflectors so that their position can be measured to millimetres when laser pulses are reflected from them. Others, like JASON-1, TOPEX/Poseidon and ENVISAT, carried radar to measure the wrinkles on the Earth's surface to an accuracy of centimetres.

The geodynamic satellites have discovered that the Earth's ocean surface is dimpled like a golf ball, with wind-driven currents piling water up into mounds against the continental shores, and warmer sea areas standing higher, in the same way that mercury rises in a thermometer. Overall the height of Earth's ocean surface is lower by about 150 metres in the north Indian Ocean (off the south coast of India) than in the western Pacific Ocean (off New Guinea); travelling eastwards from the sea off New Guinea to the sea off California is downhill by 90 metres. In the northern Atlantic Ocean, the sea off Florida is 130 metres lower than the sea off Iceland.

Additionally, the shape of Earth changes over the seasons and the years, as the mass of water covering the Earth shifts in position. Global-scale climate changes (such as the El Niño phenomenon) have melted sub-polar glaciers and changed the currents in the Southern, Pacific and Indian Oceans, causing the bulge in the Earth's equator to grow larger and mass to move away from the poles.

It is easy to imagine how the ocean surface is made irregular, because the flow of water is something that we experience on a small scale and we can extrapolate to the large scale. The irregular shape of the solid Earth is more difficult to envisage. The surface of the Earth is changed over geological time by the circulation of its inner liquid core pushing against its plastic near-surface layers. Below the surface of the Earth, the circulation of the liquid core causes plate movements that crack the continents and raise mountains. As measured by precision mapping using GPS satellites, the Tibetan Plateau in the Himalayas is rising by about 5 millimetres every year. Volcanoes grow where magma upwells from the interior, sinking when the magma is released in an eruption.

The rotation of the Earth plays its part by flattening the polar regions. And climate change affects the shape of the Earth: an ice cap grows at the South Pole and flows outwards to the edge of Antarctica, while depressed land areas released from the weight of melting glaciers spring upward. The Earth is a dynamic, almost a living, creature.

The Southern Constellations
Hidden stars revealed by the tilting Earth

> The sky is more the domain of science than of poetry.
> It is the stars as not known to science that I would
> know, the stars which the lonely traveller knows.

Henry David Thoreau, *Journal*, 1853

How did stars that cannot be seen from Europe and the Mediterranean come to have Greek and Roman names? Although the traditional constellations were named thousands of years ago, their visibility has changed dramatically since antiquity. In some parts of the world new stars have risen in the sky while ancient constellations have disappeared below the horizon. These changes are caused by a 26,000-year cycle of 'wobbles' in the Earth's rotation.

If you look up at the stars from the North Pole of the Earth, only half the sky is visible, the half that is centred around the North Pole of the sky immediately above your head (the point at the extension of the Earth's axis of rotation into the sky is called the North Celestial Pole). The other half of the sky is perpetually below the horizon, even though the Earth rotates. But if you are at the Equator, you will be able to see every part of the sky in turn at different times of the year. At intermediate European latitudes, where most of the classical constellations were named after figures from Greco-Roman mythology, half the southern sky is perpetually out of sight. Because the constellations in this part of the southern sky could not be seen from Europe or the Mediterranean, they were not named by Europeans until sailors began to explore the southern seas in the fifteenth and sixteenth centuries CE. The oldest printed star charts published in Europe date from 1515. They are woodcuts, produced in Nuremberg, Germany, the product of an innovative collaboration between the German artist Albrecht Dürer, the eminent Viennese cartographer Johannes Stabius and the German astronomer Conrad Heinfogel. The map of the southern constellations shows a remarkable 'hole' in the region never visible from the north, where no constellations had been described.

There are two peculiar exceptions to this simplified account. In one part of the sky that lies south of the bright star Formalhaut and is, in fact, readily visible from Europe, there are no constellation figures that have mythological names dating back to classical times. On the opposite side of the celestial globe is an area of the southern sky, which, although it never rises above the horizon even in southern Europe, contains a substantial part of a constellation that was known to Greek and Roman classical scholars even though they were never able to see its stars. This constellation is called Argo, the name of the ship in which, according to classical mythology, Jason sailed with the Argonauts in search of the Golden Fleece. The situation seems very puzzling: stars that were not visible from Europe came to have European classical names, while visible stars were for some reason left unlabelled. The solution to the mystery is precession.

The traditional constellations are centred on Polaris, the North Star. This star, the brightest in the constellation Ursa Minor, lies near to the centre of the 24-hour rotation of the stars, but it has not always been in 'pole position'. The axis of the rotation of the Earth shifts, cyclically, in space, wobbling like the stem of a spinning top and pointing to different stars from time to time. This wobbling is called 'precession'. The effect of precession is to twist the axis of the Earth's diurnal rotation in a circle on the sky. The Earth's axis slowly follows this circle over a period of 26,000 years. At this speed the changes in the constellations are not readily noticeable from generation to generation, but may be perceived from civilization to civilization.

In the several thousand years that have passed since the constellation figures were first recorded, precession has twisted the Earth's axis from its original position to the present North Pole. By 1515, the date of Dürer's map, the hole in the constellations in the south was quite obviously offset from the present position of the south celestial pole. This had the effect of hiding part of the sky mapped by classical astronomers, while revealing another area, which contains stars that could not be seen from Europe and the Mediterranean in classical times, and was therefore not mapped until much later. A further effect of the twist is that constellation figures (such as Orion) that used to stand erect in the sky, with their heads facing the North Pole, now lean askew, at an angle. These changes started to be apparent to astronomers as early as the second century BCE, when the Greek astronomer Hipparchus noticed discrepancies between his own observations of the sky and the constellations that had been recorded by his predecessors Eudoxus and Aratus two centuries earlier.

Precession also affects the part of the sky in which the Sun appears at the spring equinox. For roughly 2,000 years before the start of the Christian era, the Sun appeared near the constellation Aries at the spring equinox, shifting gradually into Pisces towards the beginning of the Christian era. Soon Aquarius will become the constellation of the spring equinox (some say that it already has), and this will be (or has been) the 'dawning of the Age of Aquarius'. An 'age' in this sense lasts about 2,000 years – the period of precession, 26,000 years, divided among the twelve signs of the Zodiac.

Using the known speed of precession and working backwards, it is possible to calculate that all the classical constellation figures were visible and stood erect in the sky around 2800 BCE (plus or minus 300 years) as seen from latitude 36 degrees. This corresponds to the time and place where the constellations were formulated: during the height of the great civilizations of the Tigris–Euphrates valley, which straddles this latitude.

While most of the constellations of the northern sky are named for figures from classical mythology, many southern constellation figures commemorate modern inventions, having been named by seventeenth- and eighteenth-century astronomers during expeditions to southern lands. The eighteenth-century French astronomer Nicolas-Louis de Lacaille carried out an extended astronomical expedition to the Cape of Good Hope, South Africa. He invented constellations celebrating what were at the time modern inventions such as Fornax (the Chemical Furnace) and Horologium (the Pendulum Clock). He broke down one large southern constellation, Argo Navis, into more conveniently sized component parts including the Keel, the Poop Deck and the Sails.

Known in Europe as a separate constellation only since the sixteenth century (Amerigo Vespucci claimed to be the first European to see its stars on his third voyage of 1501), the constellation of Crux, the Cross, which lies near to the South Celestial Pole, is of particular significance in Christian cultures because of its cruciform shape. In modern times the Southern Cross has become a sentimental symbol of Southern-Hemisphere patriotism – it appears on the flags of Australia, Papua New Guinea, New Zealand, Samoa and Brazil. The four stars of the Southern Cross were in fact known to Ptolemy as part of the constellation Centaurus but were 'lost' to European eyes as the Earth tilted away from the constellation due to precession. The Southern Cross ceased to be visible from the British Isles at the time that the building of Stonehenge was started (about 3000 BCE) and from northern Mediterranean shores at the time of the collapse of the Roman Empire (about 500 CE).

Because the southern stars were imagined but unseen by Europeans, and associated with exploration and long voyages, they are often used symbolically in literature and art set in the strange lands overhung by the southern skies to convey a sense of not belonging, of adventure and nineteenth-century imperialism. For example, in Thomas Hardy's poem 'Drummer Hodge' (1899), the drummer-boy killed in the Boer War and buried in the Karoo desert in the Cape lies in a grave on an isolated hill, where 'strange-eyed constellations reign / His stars eternally'.

The Sun
At the centre of the Solar System

> If there should chance to be any mathematicians
> who, ignorant in mathematics yet pretending to skill
> in that science, should dare, upon the authority of
> some passage of Scripture wrested to their purpose,
> to condemn and censure my hypothesis, I value them
> not, and scorn their inconsiderate judgement.

Nicolas Copernicus, *De revolutionibus orbium coelestium*, 1543

In the first half of the sixteenth century, the Polish cleric Nicolaus Copernicus outlined a hypothesis to replace the 1,500-year-old Ptolemaic cosmological system, proposing that all the planets, save the Moon, revolved around the Sun. Few scientific discoveries have demanded such a fundamental transformation of human thought. Yet Copernicus's revolutionary theory was actually the product of centuries of careful astronomical observation that had gradually exposed the failure of the Ptolemaic system to predict the movement of the planets.

Since the second century CE, the geocentric theory, called the Ptolemaic system, had held that the Earth was stationary at the centre of the Universe. The theory survived unchallenged for over 1,500 years as it contradicted neither literal readings of the Bible nor the immediate evidence of the five senses. Yet by the sixteenth century the Ptolemaic theory was beginning to suffer a progressive loss of confidence. It had become clear that the geocentric model required constant additions and changes to make it correspond with astronomers' actual observations. Astronomers were beginning to feel that these successive add-ons were arbitrary and did not give a decisive overall explanation for the observed behaviour of the heavenly bodies. The most serious problem was in the addition of epicycles to explain why some planets did not appear to orbit the Earth as predicted by the Ptolemaic theory.

Because the Earth has a shorter orbit than the outer planets (Mars, Jupiter, Saturn, Uranus and Neptune), it overtakes them from time to time as it revolves around the Sun. When viewed from the Earth on these occasions, the outer planets seem to halt in their orbits and move backwards in a loop. This is called 'retrograde motion'. As first proposed by Appolonius of Perga (*c.* 200 BCE) and Hipparchus (*c.* 130 BCE), Ptolemy added epicycles to his system to account for this loop.

An epicycle was envisaged as a kind of revolving wheel that carried the planet on its outer rim, while itself revolving in an orbit around the Earth. The combination of the two motions was thought to produce the retrograde loop. In modern engineering, the concept of epicycles survives in the name of so-called 'planetary gears', in which a small gear wheel rotates outside a central one, all enclosed within a hollow toothed chamber.

Despite this attempt to refine the existing Ptolemaic theory, discrepancies between the predicted and observed orbits of the planets continued to build up over time. This was particularly a problem with the planet Mars, which moves unusually quickly through an especially eccentric orbit, and whose orbit seen from Earth is therefore more than usually complex. To predict the movement of Mars accurately, astronomers were obliged to add ever more epicycles and to assign these epicyclical 'wheels' a variety of different sizes and spin rates.

Although the theory of epicycles eventually did allow astronomers to calculate the movements of the planets with relative accuracy – no inconsiderable achievement – the epicycles did not seem to be grounded in real observation. All they did was make the calculations work. Moreover, the theory never seemed to be reaching a stage where it was able to provide a satisfactory unified explanation for the motions of the planets. When further epicycles had to be added, it seemed that there was no end to the arbitrariness of the theory ('wheels within wheels' is an expression recognizing the arbitrariness and secrecy of complex social machinery, like the astrologers' systems of epicycles, and resonating with the biblical expression in Ezekiel 1:16).

The situation began to change in 1543, when the Polish cleric Nicolaus Copernicus published his *De revolutionibus orbium coelestium* ('On the Revolutions of the Heavenly Spheres'), proposing a radical new hypothesis to replace the Ptolemaic theory: all the planets, save the Moon, revolved around the Sun in a series of concentric orbits. Copernicus produced a remarkably accurate diagram of the Solar System that showed the order of the planets from the Sun, which he fixed by making a smooth progression from the Sun outwards of the orbital speed of each planet, with the slowest furthest from the Sun. Copernicus's theory also simplified the calculations of planetary motions, explaining retrograde motion as a visual effect created by the orbital movement of the Earth, although he still formulated the calculations in terms of a (smaller) number of epicycles. These were astonishing thoughts – too astonishing for many, and Copernicus's hypothesis remained just that for more than sixty years, until it was given convincing form by the brilliant and eccentric German astronomer Johannes Kepler.

Kepler based his calculations on observations by his tutor, Tycho Brahe. Brahe was a sixteenth-century Danish astronomer. A strong Protestant, Brahe did not believe that the Earth moved, because that appeared to contradict biblical texts. Brahe had proposed that the Moon and Sun orbited the Earth, but that all the other planets revolved around the Sun. However, the accuracy of Brahe's observations provided the data that allowed his pupil to complete the simplification of the Copernican theory, which, ironically, was anathema to Brahe's literalist religious beliefs.

In his *Astronomia Nova* ('New Astronomy', 1609) Kepler compared the orbit of Mars in the heliocentric Copernican planetary system, the geocentric Ptolemaic system and Tycho's system (a compromise between the former two) – all using variations on the idea that celestial orbits were circular. On the basis of Tycho's own data, Kepler showed that it was more accurate to use the Copernican theory to calculate the position of Mars, but only when its orbit was assumed to be an ellipse, not a circle. This was not only a more accurate basis for calculation, it was a simpler, but novel, model of how the planets moved.

In simplifying Copernicus's theory, Kepler produced the first recognizably modern plan of the Solar System. He abolished the concept of epicycles altogether, and in their place offered his own description of the motions of the planets. He calculated tables to predict where the planets would be in the future (the *Rudolphine Tables*, published in 1627), and was able satisfactorily to predict that Mercury would be aligned with the Sun in 1631 so accurately that it would pass in front of the Sun's disc. This 'transit of Mercury' was witnessed by the French astronomer Pierre Gassendi. Eight years later, using Kepler's tables and theories, the English cleric Jeremiah Horrocks calculated a transit of Venus and, with the merchant William Crabtree, actually saw the transit with his own eyes, just before sunset on 24 November 1639 (plate II). They were both ecstatic with their discovery: Horrocks wrote that, 'rapt in contemplation, [Crabtree] stood for some time motionless, scarcely trusting his own senses, through excess of joy; for we astronomers are of a womanish disposition and are overjoyed with trifles…'.

These observations were convincing proof that Kepler's calculations of the planets were the most accurate ever produced. How did he do it? Kepler showed that the orbits of the planets about the Sun, particularly the orbit of Mars, were simple ellipses. He also demonstrated that there were basic relationships between the motions of the planets and the sizes of their orbits – his 'laws' of planetary motion. Not only did these interrelationships provide a grip on the calculations, they removed a lot of the arbitrariness in the arrangement of the planets. Kepler attempted an explanation of these relationships by proposing that there existed a 'magnetic' virtue or force between the Sun and the planets, including the Earth and Moon. In so doing, he prepared the way for the identification of the force of gravity by Isaac Newton.

The theory of epicycles had seemed arbitrary to astronomers, but the implications of the Copernican system were nothing short of startling. Far from lying stationary at the centre of the Universe, the Earth is moving. Our planet spins on its axis, it orbits around the Sun, and it wobbles as it orbits. Despite the lack of sensation its gyrations produce in us, the Earth's motions are dizzying as a dance.

DISCOVERING

THE SOLAR

SYSTEM

The Orbits of Comets
Disasters, sun-grazers and the 'Lady's Comet'

...Now we know
the sharply veering ways of comets, once
A source of dread, no longer do we quail
Beneath the appearance of bearded stars.

Edmond Halley, Dedication to Newton's *Principia*, 1686

Unlike stars and planets, comets appear without warning and move rapidly and erratically across the sky. In early cultures, comets were regarded as harbingers of doom because of this behaviour, but in the centuries following the Enlightenment, this same unpredictability has become an irresistible attraction for comet-hunters. One of the earliest and most famous comet-hunters was an eighteenth-century English woman, Caroline Herschel, who laboured under extreme conditions to discover fourteen comets and was rewarded with a royal stipend for her pioneering efforts. She described in her memoirs her first discovery and the change of status it earned her:

On the 1st of August 1786 I found an object very much resembling in colour and brightness the 27th nebula of Messier's catalogue, with the difference however of being round. I suspected it to be a comet, but, a haziness coming on, it was not possible to entirely satisfy myself as to its motion until the following evening, and thus to confirm it as my first comet discovery. My brother William was on his return commanded to show it to the King, who said that it was very small and had nothing striking in its appearance. Miss Fanny Burney acclaimed my comet as the first 'lady's comet'. It gave great pleasure to the ladies of the Court. I heard that Princess Augusta was in particular desirous that the lady guests should view it, calling them from the card-table.

Partly as a result of this, and in specific consequence of a recommendation made to the King by Sir Joseph Banks, President of the Royal Society, a salary of 50 pounds per year was settled on me as an assistant to my Brother. In October 1787 I received the first quarterly instalment of 12 pounds 10 shillings. It was the first money in all my lifetime that, at the age of 37 years, I ever thought myself at liberty to spend to my own liking.

Comets are small, dark bodies in the Solar System, which are hard to see when far from the Sun. They have a solid part, the 'nucleus,' which is composed of both ices and solid material (plate IX). When a comet comes close to the Sun, its nucleus melts, vaporizes and becomes 'active'. It develops a bright, dusty atmosphere, the 'coma', and 'tails' of gas and dust, which reflect more sunlight (plate XII). Coupled with the speed of the comet as it approaches the Sun, this means that comets can spring into sight, unexpectedly and suddenly.

For a long time, comets were interpreted superstitiously because, in an age when the cyclical movements of the stars were thought to control events on Earth, they arrived sporadically and unpredictably. In his 1665 book *De Cometis*, English astrologer John Gadbury warned that comets were 'threatening the world with Famine, Plague and War: To Princes, Death! To Kingdoms, many Crosses; To all Estates, thunder, lightning and earthquakes, inevitable Losses! To Herdsmen, Rot; to Plowmen, hapless Seasons; To Sailors, Storms, To Cities, Civil Treasons!' The word 'disaster' to designate events like these is a linguistic fossil left over from this era, meaning literally 'bad things from the stars'; likewise 'influenza' is a disease induced, it was believed, by the influence of comets.

Anyone who is looking in the right place at the right time can discover a comet, which will usually be named for its discoverer.

The most successful comet discoverers are of course those who dedicate lots of time to systematic searches – one highly successful Japanese comet-hunter is said to have dedicated the rest of his lifetime to finding comets after a deathbed vow to his dying father. The usual method is to sweep a telescope or binoculars from side to side to examine a section of the sky, looking for objects that are fuzzy. Nebulae, galaxies and star clusters can be eliminated by reference to a catalogue or atlas. French comet-hunter Charles Messier compiled a list of such objects that might be mistaken for a comet. Published between 1774 and 1781, this became known as the Messier Catalogue, and is still in use. The clinching difference is that a comet moves in the sky, whereas nebulae and galaxies are stationary.

Armchair astronomers have discovered over 3,000 comets using data from the Solar and Heliospheric Observatory (SOHO) satellite; the satellite was launched late in 1995 and has been gazing constantly at the Sun for more than twenty years. Comets show in SOHO's camera as they graze the surface of the Sun, some of them approaching the Sun but not seen to leave (i.e. melting). Most of them are members of the Kreutz sungrazing family. In 1888, Heinrich Kreutz discovered that a number of comets have similar orbits passing near the Sun. They come from a single comet that has disrupted into many fragments, which have further broken into small pieces about 10 metres in size – the comets seen in SOHO's cameras. Armchair astronomers use online archives to scan SOHO pictures day by day for new discoveries of these fragments before they disappear forever.

The Satellites of Jupiter

Shattering the crystal spheres

... behold! afar,
Four radiant Moons surround th' imperial Star,
Large as our boasted World; whole silver Light
His regions visit in the Gloom of Night;
Nor this the Fancy of deluded Eyes;
Mark'd are their Periods thro' sublimer skies:
Oft does th' Astronomer with his Tube display,
And view 'em in Eclipse with pleas'd Survey;
To this curious Genius Knowledge owes,
Of Light's swift Motion, and its Measure knows.

Moses Browne, *Sunday Thoughts: The Works and Rest of the Creation*, 1752

The belief that the heavens orbited the Earth on a series of crystal spheres had survived for 2,000 years, but shortly after the invention of the telescope it shattered in an instant. A few years after Kepler and Copernicus had used their calculations of planetary movements to develop the theory that the Earth and the planets orbited the Sun, Galileo Galilei saw the irrefutable evidence with his own eyes: mountains on the Moon, and four mysterious stars near Jupiter.

In 1608 Hans Lippershey (or Lipperhey), a German-born optician living in Middelburg, Zeeland, applied for a patent from the Dutch government for 'a certain device by means of which all things at a very great distance can be seen as if they were nearby, by looking through glasses which he claims to be a new invention'. One story is that Lippershey discovered the principle of the telescope when two children were playing with lenses in his shop and looked through two at the same time, one held up behind the other, exclaiming with surprise at what they saw.

In Venice later that year the learned monk Pietro Sarpi, recently retired from high office in the Venetian government, heard about the patent application, and when a few months later Sarpi was visited by a protégé, Galileo Galilei, they discussed Lippershey's invention. Galileo was professor of mathematics at Padua University in the Venetian republic and he realized the importance of the invention to Venice, a maritime power. He had been seeking a salary increase and in 1609 he made a prototype of a telescope that, through Sarpi, he brought to the attention of the Venetian authorities. Galileo was able to describe ships approaching the port of Venice before they could be seen at all with the unaided eye. The value of the invention was clear and his salary was doubled, although there were strings attached to his promotion: he would never again get a salary increase, and could never move from Padua. This resulted in him looking elsewhere for a new position, a prize that the momentous astronomical discoveries he was about to make with his telescope would eventually win for him.

In the winter of 1609–10 Galileo observed the Moon and saw bright spots in the shadowed part of its surface that gradually grew in size and merged with the illuminated area as the month progressed. He correctly interpreted the bright spots as mountaintops that had caught the first rays of the Sun, moving into full sunlight as the Moon rotated. He measured the heights of these lunar mountains using their shadows, calculating one at 6 kilometres high. He saw that some mountains were arranged in straight mountain ranges and others in circles, surrounding craters (plate VIII). He discovered that the Moon was not a smooth, perfect sphere, as taught by Aristotle and Ptolemy; rather, its surface was 'rough and uneven, and just like the surface of the Earth itself'.

In England, a mathematician and cartographer, Thomas Harriott, used a telescope to observe the Moon and drew it in August 1609, several months before Galileo. He also discovered sunspots, but he never published his astronomical work and had no influence on the development of science whatsoever. He thus remains largely unknown for this achievement. He is, however, remembered as one of the founders of the Virginia colonies, having visited Roanoke Island in 1585–86 and written an influential report about its agricultural and mineralogical potential, and its Algonquian inhabitants.

On 7 January 1610 Galileo drafted a letter to an unknown recipient describing a further momentous discovery that he had made the previous night: 'And besides my observations of the Moon, I have observed the following in other stars. First that many fixed stars are seen with the telescope, which are not otherwise discerned; and only this evening, I have seen Jupiter accompanied by three fixed stars, totally invisible by their smallness....'

At first Galileo did not think there was anything remarkable about these stars: a triplet arranged in a straight line through Jupiter, two on one side and one on the other. But, according to his observation journal, when he came to look at Jupiter again on the 8th of January the three stars were all on the other side of Jupiter. Presumably Jupiter had moved from its previous position. The 9th was cloudy, and on the 10th and 11th there were two stars only, on one side of Jupiter, with the third conjoined with Jupiter (or so Galileo speculated). On the 12th the three stars were again arranged differently: two on one side of Jupiter, one on the other. 'It appears that around Jupiter there are three moving stars invisible to everyone up to this time.' On the 13th, Galileo realized that there were in fact four little stars; he saw them all again on the 15th. Galileo seems originally to have thought that the stars were moving back and forth in a straight line. But if this was the case, how did they pass one another? Suddenly Galileo realized that the four 'stars' were actually in orbit around Jupiter. In an instant, the four tiny stars had disproved the 2,000-year-old Ptolemaic theory that every celestial body orbited around the Earth.

The orbital motion of the satellites around Jupiter was very like the orbital motion of the planets around the Sun as expressed in the Copernican theory, and also very like the motion of the Moon around the Earth. It became clear that the Earth was just like the other celestial bodies, and that it was not the centre of the Universe.

Galileo wrote up his discoveries in January and February 1610 in a book called *Sidereus Nuncius* ('Starry Messenger'). He dedicated the book to Cosimo II de' Medici. The Medici family was a powerful and wealthy family that ruled Florence from the thirteenth to the seventeenth centuries. Galileo had tutored Cosimo in mathematics, and at the age of nineteen, in 1609, the young man succeeded to the title of Grand Duke of Tuscany on the death of his father, Ferdinand. To flatter the family, Galileo christened the satellites of Jupiter 'the Medicean stars', although this term did not stick. As a reward and as a mark of esteem, in 1610 Cosimo appointed Galileo as his Philosopher and Mathematician for his lifetime, and Chief Mathematician at the University of Pisa.

The moons and their shadows pass in front of one another and provide a spectacular series of occultations and eclipses. Sometimes the eclipses appear to run ahead and then behind schedule. The English Reverend Moses Browne drew on the Danish astronomer Ole Rømer's interpretion of this phenomenon in his poem, at the head of this chapter. Rømer realized that the variations had to do with the speed of the light coming from Jupiter to Earth over a distance that changed as the planets revolved in their orbits. His measurements of these intervals enabled him to make the first calculation of the speed of light in 1676. The fact that light travelled with a velocity was a discovery with far-reaching consequences, leading eventually to Einstein's theory of relativity.

In the space age, Jupiter's satellites can be seen as individual worlds, each with its own different character (plate III). Io is pockmarked with volcanoes and scoured by lava flows. Europa is completely covered with cracked floes of ice. Ganymede is made of a mixture, half and half, of rock and ice, its surface plastic, mobile and wrinkled under tectonic forces. Callisto is heavily cratered, like Ganymede, and indeed, both look very like our own Moon.

The Phases of Venus
Revealing the shape of the Copernican System

In questions of science the authority of a thousand is not worth the humble reasoning of a single individual.

Attributed to Galileo by François Arago in his *Eulogy to Galileo*, 1874

Having discovered that the 'Medicean Stars' orbited Jupiter rather than the Earth, Galileo turned his telescope towards the planet Venus. He discovered that Venus had phases similar to the Moon, which proved that Venus was orbiting the Sun and that Copernicus's theory of the Solar System was correct. Galileo's discoveries were at first acclaimed by astronomers, but then perceived as a threat to the biblical worldview, earning him persecution from the Church during his lifetime. History now recognizes Galileo as one of the world's great scientists.

In 1610, in Pisa, after discovering the satellites of Jupiter with his new telescopes in Venice, Galileo was finally able to observe the planet Venus – when he had first assembled his telescope, Venus had been too near to the Sun to be examined closely. He saw that Venus had phases like the Moon. When the planet was at its greatest distance from the Sun, it looked like a half-moon, with the bright side facing in the direction of the Sun. As it approached the Sun again, its phase either increased towards full circular illumination, or narrowed to a thin crescent. Galileo reported his discovery using a Latin anagram, which he sent to Kepler. The coded sentence (in not very elegant Latin) was '*Haec immatura a me iam frustra leguntur o.y.*' It can be translated as 'Things not ripe for disclosure are read by me', but its letters can be rearranged to read '*Cynthiae figuras æmulatur mater amorum*'. This means 'The shapes of Cynthia [the Moon] are emulated by the mother of loves [Venus]'.

The anagram method of making a coded announcement was a device used in the seventeenth century to establish priority for a discovery – if the announcement was made straightforwardly and was promulgated at the slow pace of communication of the time, someone who got the announcement earlier than anyone else could falsely claim the discovery as his own, and pretend surprise when the announcement became more widely disseminated. The terse nature of the anagram also bought time and room for manoeuvre for the discoverer as a hedge against an incomplete or misinterpreted discovery.

Galileo's discovery of the phases of Venus showed immediately that Venus travelled around the Sun in an orbit that lay inside the orbit of the Earth. When Venus was on the far side of the Sun, its face was fully illuminated, like the Full Moon. When Venus was moving between the Earth and Sun, its unilluminated rear partially faced the Earth and it showed only a thin crescent of light. This geometry was exactly the arrangement that Copernicus had hypothesized in his model of the Solar System, and could not be reconciled with the Ptolemaic theory that Venus orbited the Earth in a crystal sphere between the Earth and the Sun.

Galileo's observations also confirmed another of Copernicus's predictions: that the size of Venus would appear to change as the planet came closer to Earth and got further away. Galileo saw that Venus did indeed change size – it appeared smallest when it was at its most distant, beyond the Sun, and showing a full face, and four times larger when it was crescent-shaped and at its closest approach to Earth.

As Galileo himself realized, this discovery was crucial in confirming Copernicus's theory of the Solar System. In principle, the phases of Venus were still consistent with Tycho Brahe's compromise between the Copernican theory and the Ptolemaic theory. Brahe had proposed in 1583 that the planets orbit the Sun, which itself orbits the stationary Earth. However, Galileo dismissed Brahe's theory (also called the 'Tychonic theory') because it assumed that the Sun caused the planets to move, while leaving the Earth stationary and unaffected, even though some planets were evidently bigger than the Earth.

Galileo knew that his discovery would not be readily accepted. On New Year's Day in 1611 he sent to the Tuscan ambassador in Prague the solution to the Venus anagram, explaining that the phases meant that the planet must orbit around the Sun: '…something indeed believed by the Pythagoreans, Copernicus, Kepler and myself, but not proved as it is now. Hence Kepler and other Copernicans may glory in their successful theories, although as a result we will be thought to be fools by most bookish philosophers, who will regard us as men of little understanding or common sense.'

The consequences for Galileo were actually much worse than this. His discoveries were acclaimed by astronomers and even sophisticated churchmen like Cardinal Maffeo Barberini (later Pope Urban VIII), but he was denounced to the Inquisition as a heretic by Tommaso Caccini, a Dominican friar, and cautioned by Cardinal Bellarmine to treat the Copernican theory as a hypothesis only. The telescope was derided as a device capable of making things appear in the sky that were not actually there. (Galileo said that he would pay 10,000 *scudi* – ten times his annual salary – to anyone who could make a telescope that showed satellites around one planet but not the others.) He was told in 1616 not to advocate or teach Copernican astronomy except as a hypothesis, but nevertheless continued to publish his scientific work, and evidently pushed the boundaries of church tolerance too far. Galileo was put on trial in Rome for teaching, contrary to Scripture, that the Earth moved. In 1633 he was convicted and forced to recant:

I, Galileo, son of the late Vincenzio Galilei, Florentine, aged seventy years, arraigned personally before this tribunal and kneeling before you…swear that I have always believed, do believe, and by God's help will in the future believe all that is held, preached, and taught by the holy Catholic and apostolic Church....I have been pronounced by the holy office to be vehemently suspected of heresy, that is to say, of having held and believed that the Sun is the centre of the world and immovable and that the Earth is not the centre and moves....I abjure, curse, detest the aforesaid errors and heresies...and I swear that in future I will never again say or assert, verbally or in writing, anything that might furnish occasion for a similar suspicion regarding me.

He was thus banned from publishing any future scientific discoveries and placed under house arrest until his death in 1642.

Uranus
The first new planet

It has generally been supposed that it was a lucky accident
that brought the planet to my view; this is an evident
mistake. In the regular manner that I examined every
star of the heavens, not only of that brightness but many
far inferior, it was that night its turn to be discovered. I
had gradually perused the Great volume of the author of
Nature, and was now come to the page which contained a
seventh Planet. Had business prevented me that evening
I must have found it the next, and the goodness of my
telescope was such that I perceived its visible planetary
disc as soon as I looked at it. And by the application of my
micrometer I determined its motion in a few hours.

William Herschel on the discovery of Uranus

The only planets that were known to the astronomers of antiquity
were those that could easily be seen with the naked eye: Mercury,
Venus, Mars, Jupiter and Saturn. This changed in the eighteenth
century when, armed with a powerful home-made telescope, a
music teacher from Bath, England, discovered Uranus, the first
new planet identified in recorded history, and the third-largest in
the Solar System.

William Herschel was born in 1738 in Hanover, at a time when
it was a British possession. He followed his father into a career as
an oboeist in the band of the Hanoverian Guards. They fought as
part of the British army at the battle of Hastenbeck in 1757, after
which the younger Herschel escaped to England, eventually settling
in Bath. He enjoyed great success as a music teacher to society
ladies and as a concert artist. He was eventually joined in Bath by

his sister, Caroline, who not only kept house for him and tried to defend him against predatory widows but also accompanied him, singing, in his concerts. William's father had been interested in astronomy and William also studied the subject, forming an ambition to see the heavens with his own eyes. He made telescopes, casting and grinding the mirrors himself in the basement of his house, designing and forming the tubes from wood and tin, and erecting the telescopes on the garden lawns of his various houses. Herschel then set out systematically to 'review' the entire sky with his tele- scope, inspecting every star and even the spaces between them.

On 13 March 1781 he discovered a 'curious either nebulous star or perhaps a comet'. Tracking it over the next hours and days, he found that it moved and could not be a star. Over the next few weeks, the curious object was also observed by astronomers at Greenwich and Oxford, and proved to be in a near-circular orbit outside Saturn's, rather than the highly eccentric orbit crossing the Solar System that would be expected of a comet. Furthermore, there was no trace of a fuzzy coma or tail, as a comet would have, and the object had a circular disc like a planet. It was, indeed, a planet – the first discovered since antiquity. As a planet of sixth magnitude in brightness, Uranus is marginally visible with the unaided eye and easily seen in modest telescopes. It was mistakenly recorded several times as a star before Herschel's discovery.

Herschel was invited to London to tell King George III of the new planet. He was asked to erect a telescope at Windsor Castle in order to show astronomical sights to the royal household. He was given a patronage appointment as the King's Royal Astronomer and given a stipend – his sister Caroline, who assisted him with his observations and discovered the first comet found by a woman, was later given a stipend too, of half as much. In recognition of this royal patronage, William Herschel suggested that the new planet should be named after the era of its discovery 'in the reign of King George III' as 'Georgium Sidus' – the Georgian star (or planet). This name was used for a time in England, but never in any other country, and it was the suggestion of the mythological name Uranus by the German astronomer Johann Bode that eventually stuck.

Uranus is the third-largest of the planets in the Solar System, and ranks fourth in mass. It has a characteristic blue-green appearance due to a high layer of clouds of methane ice. It has a ring system, discovered in 1977 when high-speed photometers were used to observe a star that, by chance, was being occulted (passed in front of) by Uranus, with the intention to study how the star's light faded in the planet's atmosphere. The starlight was unexpectedly blocked by the unknown rings. The rings were first imaged in 1986 by the Voyager 2 spacecraft.

Uranus has five bright satellites and more than a dozen fainter ones. The satellites provided the first evidence of the unusual tilt of Uranus. They orbit around Uranus's equator and show that its pole is tilted by more than 90 degrees. One theory is that Uranus originally had a more normal tilt, but that, late in its formation process, it was struck by an Earth-sized planetesimal that knocked it over. Because its rotational axis stays fixed in direction while Uranus orbits around the Sun, its polar regions are sometimes pointing directly towards the Sun and sometimes at right angles to it. By contrast, the Earth's polar regions are never presented directly to the Sun, although they do tilt a little because the Earth's axis is tilted by 23½ degrees to its orbit. This small tilt is the reason for the pattern of the Earth's seasons. The differences in solar heating during Uranus's orbital 'year' are much more extreme than at Earth, and its seasons are peculiar – for example, during its summer time, when the Sun is always above its horizon, one of its polar regions will be hotter than the equatorial regions ever are.

Neptune
The planet discovered by the pen

The method pursued by M. Le Verrier totally differs from all previous attempts of geometers and astronomers. The latter have sometimes accidentally found a movable point, a planet, in the field of their telescopes. M. Le Verrier perceived the new body without the necessity of casting a single look towards the heavens. He saw it at the end of his pen; he determined by the mere force of calculation the place and approximate magnitude of a body situated far beyond the hitherto known limits of our planetary system, of a body the distance of which from the sun exceeds 2,800 millions of miles, and which, seen in our powerful telescopes, barely exhibits a sensible disc.

François Arago, *Comptes Rendu*, 1846

While Uranus was discovered with a telescope, Neptune was discovered with a pen and paper – twice. Two very different astronomers – the Frenchman Urbain Le Verrier and the Englishman John Couch Adams – independently tried to explain why Uranus was being pulled off course in its orbit. Both sets of calculations pointed to a previously unknown planet, making Neptune the first planet to be located through mathematics rather than by direct observation of the heavens.

After the discovery of Uranus by William Herschel in 1781, several previous observations of the planet had been found – made and recorded by astronomers who had not recognized it as a planet. By 1830 these records enabled an accurate orbit to be calculated for Uranus. At the same time, it became clear that Uranus was departing from this orbit. The Director of the Paris Observatory, François Arago, suggested to his colleague Urbain Le Verrier that a previously unseen planet was pulling the planet off track. In 1845 Le Verrier calculated the expected position. He used Bode's Law to assume the distance of Neptune from the Sun, although the accuracy of Bode's Law actually breaks down after Uranus. Le Verrier sent his prediction to Johann Galle, an astronomer at the Berlin Observatory. Galle, together with his assistant Heindrich D'Arrest, began a search on the same night that they received the letter, 23 September 1846.

At D'Arrest's suggestion, Galle used the latest star chart of the area, which had only just been produced. Within thirty minutes they had identified a star that was not on the map. They confirmed that it was the new planet on the following night by its motion relative to the other stars. Galle wrote to Le Verrier, saying, 'Monsieur, the planet of which you indicated the position really exists'. Le Verrier replied, 'I thank you for the alacrity with which you applied my instructions. We are thereby, thanks to you, definitely in possession of a new world.'

Meanwhile, in England, the Scottish mathematician Mary Somerville had suggested to a young Cambridge student, John Couch Adams, that an unknown planet was affecting the orbit of Uranus. With a letter of introduction from James Challis, Director of the Cambridge Observatory, who had been impressed by Adams's preliminary calculations regarding the position of the hypothetical planet, Adams applied to George Airy, the Astronomer Royal, for research assistance. However, due to his youth, humble background and reticent manner – and perhaps also due to Airy's unapproachable character – Adams twice failed to secure an interview with Airy. Airy dismissed the discovery of a new planet as not the job of the Greenwich Observatory, and told none of his colleagues about Adam's prediction. He passed the project back to Challis, who began a somewhat half-hearted search for Neptune on the basis of Adams's calculations. But Challis did not have the same up-to-date charts as Galle, and had to compare successive observations of the same area of the sky to see if there was a star that moved.

Hobbled by a rigid institutional hierarchy and outdated materials, Adams and Challis had been overtaken by the brisk efficiency of Le Verrier and Galle, although Adams's calculations, which he shared only with Challis and Airy, had accurately predicted the position of Neptune eight months before Le Verrier's findings were published. When Airy and Challis were forced to justify their delay in following up Adams's calculations, Adams is said to have reacted with characteristic modesty, offering his congratulations to Le Verrier without a trace of bitterness – as the president of the Royal Astronomical Society it was he who handed its gold medal to Le Verrier in 1875. Eventually both Adams and Le Verrier would be credited jointly with the discovery of Neptune.

After Neptune had been located, William Lassell, a wealthy Liverpool brewer and amateur astronomer, was one of the first to inspect the new planet with a large telescope. According to family lore, Lassell just missed out on being the discoverer of Neptune, a letter communicating Adams's prediction having been accidentally destroyed by an overzealous maid clearing away and burning rubbish. When the coordinates of 'Le Verrier's planet' were published in *The Times*, Lassell rushed to examine its position in the sky with his telescope. He saw that Neptune had a distinct disc, confirming its identification as a planet. During the same observation, he also discovered Neptune's satellite, Triton. Triton orbits Neptune backwards, and at a steep angle, suggesting that it was not formed together with Neptune, but was a passing planet that Neptune captured.

The naming of Neptune generated as much controversy as its discovery. In keeping with the custom for naming celestial features after Roman mythological characters, Galle suggested that the new planet should be called 'Janus', and Challis proposed 'Oceanus'. Le Verrier, eager to put his own stamp on 'the planet exterior to Uranus', first proposed 'Neptune' and later 'Le Verrier', but in the ensuing dispute over who should be credited with the planet's discovery, the latter suggestion received fierce opposition outside of France. By the end of the year the more neutral 'Neptune', the name of the Roman god of the sea, had become the accepted usage.

Neptune is the outermost of the four giant planets, forming a pair with Uranus. It rotates quickly and is slightly flat (or 'oblate'). Curiously it emits more than twice as much energy as it receives from the distant Sun. The excess energy is generated by the cooling of Neptune's hot interior. Neptune and its half-dozen satellites were explored by the Voyager 2 fly-by in August 1989, which discovered Neptune's 'Great Dark Spot', thought to be a hole in the methane atmosphere large enough to fit the entire Earth. Voyager also obtained images of Triton. They showed a bright south polar cap. North of the polar cap, a rugged terrain, reminiscent of the skin of a cantaloupe melon, is cross-cut by a pattern of intercepting ridges, left from some past tectonic event.

Asteroids
Remnants of the early Solar System

> The discovery of a new dish does more for the happiness
> of the human race than the discovery of a star.

Gastronome and amateur scientist Jean Anthelme Brillat-Savarin, *The Physiology of Taste: Or, Meditations on Transcendental Gastronomy*, 1825

In the sixteenth century, the German astronomer Johannes Kepler noticed that there was a rather large gap in the arrangement of the planets in the Solar System. This mysterious gap between Mars and Jupiter led to the discovery of asteroids. Some asteroids are very small planets, as thought when they were first identified in the nineteenth century, and others fragments from the collisions of larger asteroids, but many are actually scrap material left over from the formation of the planets in the early Solar System (plate VII). Some are spherical, but most are random in size and shape. Most follow orbits in a belt between Mars and Jupiter.

Setting the Sun–Earth distance at 10 units, the distances from the Sun to the known planets were as follows:

Sun–Mercury	3.9 units
Sun–Venus	7.2 units
Sun–Earth	10 units
Sun–Mars	15 units
Sun–Jupiter	52 units
Sun–Saturn	96 units

Kepler saw that the distances between the planets roughly doubled at each step. Yet Jupiter was nearly four times the distance of Mars from the Sun – twice the expected amount. Kepler mused whether there might be an undiscovered planet in the gap. Isaac Newton noticed the same phenomenon and suggested that perhaps Providence had put Jupiter and Saturn at an extra-large distance from the Sun to minimize their disruptive effects on the inner planets of the Solar System.

In 1766 Johann Daniel Titius, a professor of physics at the University of Wittenberg, discovered a more accurate formula for estimating the distances of the planets from the Sun. The formula was popularized by a German astronomer, Johann Bode, and consequently became known as Bode's Law (more correctly, but infrequently, called the Titius–Bode Law). The formula is still regarded as an interesting fact, although in the intervening 200 years no one has been able to explain why the planets in the Solar System are spaced apart so regularly.

For all planets except Mercury, the formula for Bode's Law works like this:

$$A = 4 + 3 \times 2^n$$

A represents the distance from the planet in question to the Sun (measured in units equivalent to $\frac{1}{10}$ the distance between the Earth and the Sun). Mercury has a fixed value of $A = 4$. n represents the consecutive order of the planets after Mercury in the Solar System, beginning with Venus at 0: 0, 1, 2, 3…

Applying Bode's Law to all the known planets gave the following results:

	Actual distance (as measured)	n	Distance according to Bode's Law
Sun–Mercury	3.9 units		4 + 0 = 4 units
Sun–Venus	7.2	0	4 + 3 = 7
Sun–Earth	10	1	4 + 6 = 10
Sun–Mars	15	2	4 + 12 = 16
?	?	3	4 + 24 = 28
Sun–Jupiter	52	4	4 + 48 = 52
Sun–Saturn	96	5	4 + 96 = 100
Sun–Uranus	192	6	4 + 192 = 196

If you compare the 'actual' distance of the planets from the Sun with the figures calculated by Bode's law, you can see that the law produces a good (but not exact) estimate for all the planets known to astronomers up to the eighteenth century (Mercury–Saturn).

There are, however, some problems. First of all, there is the funny way that the distance to Mercury is calculated. The value for Mercury is really just put in to make the formula look like it works. Secondly, there is an obvious gap at $n = 3$ between Mars and Jupiter. Like Kepler, Bode thought that there must be an undiscovered planet in the gap: 'Can one believe that the Creator of the Universe has left this position empty? Certainly not!' When William Herschel discovered the planet Uranus in 1781, belief in the validity of Bode's Law strengthened, because it fitted the formula so well. But the gap became even more significant. It was an obvious challenge to discover the unseen planet.

The court astronomer of the Duchy of Saxe-Gotha, Germany, Baron Franz Xaver von Zach, took up the search for the new planet in September 1800 and organized a team of two dozen astronomers to share the work. They became known as the 'Celestial Police'. The first day of the new century (properly reckoned), 1 January 1801, brought success even before the celestial police could get to work. A member of the team (who did not even know that his efforts had been volunteered!) discovered the new planet.

The successful astronomer was a Sicilian monk, Father Giuseppe Piazzi, who came across a moving object as he constructed a star catalogue with a telescope in Palermo. As his observations progressed, it became clear that the new object was not a comet. It had a nearly circular orbit in the right zone between Mars and Jupiter, with a distance of 28 units as measured by Bode's Law. Piazzi named the planet Ceres, after the Roman goddess of the harvest and the patron goddess of Sicily, but subsequently lost sight of the object, the sequence of his observations interrupted by illness and the passage of the asteroid out of sight behind the Sun. However, the German mathematician Carl Friedrich Gauss was able to compute its orbit, which enabled Piazzi's planet to be relocated.

When William Herschel examined the new planet with his large telescope, he could see no disc. The planet must be small – a 'minor planet' – and Herschel used a new word to describe it: 'asteroid', meaning 'an almost star-like object'. (The word was suggested by Herschel's son.) But then, to everyone's surprise, another minor planet was discovered in 1802 by the German Heinrich Wilhelm Matthias Olbers, who was a doctor in Bremen by day, and an amateur astronomer by night. Olbers located a second asteroid in 1807. A fellow German astronomer, Karl Ludwig Harding, had discovered yet another asteroid in 1804. Like Ceres, these three asteroids were named after classical goddesses: Pallas, Vesta and Juno. A further hundred new asteroids had been discovered by 1868, two hundred by 1879 and three hundred by 1890 – as astrophotography became more sensitive, so many asteroids began to spoil photographs of the stars that by the late nineteenth century they were dismissed as 'vermin of the skies'.

The discovery of asteroids went through four phases. The first phase consisted of targeted searches by astronomers who had reason to believe there was something in the gap between Mars and Jupiter; these searches produced Ceres, Pallas, Vesta and Juno. After it became clear that there were many such asteroids, there were systematic searches by eye and by photography to sample the population. This produced thousands more. Most astronomers got bored, and searches were taken up by amateur enthusiasts caught up in the romance of discovering and naming a new world, however minor. Then it became clear that there was the potential for an asteroid to hit the Earth and cause damage, and that it might be possible to mitigate against this possibility by some sort of space intervention – pushing an asteroid found to be a threat onto a new course so it misses us, or blowing it up. NASA got involved and led searches using modern imaging techniques and computing power to process the data. In this most recent, fourth phase, still in progress, asteroids are being found by the thousands, discovering the smaller members of the population. NASA flags up the asteroids that have orbits such that they have the potential to collide with Earth, so-called 'Near-Earth Asteroids'.

Compared to the planets of the Solar System, most asteroids are very small. It is estimated that there are 200 asteroids over 100 kilometres across, 1,000 over 30 kilometres, and perhaps 25 million over 100 metres. Presumably there are even bigger numbers of asteroids in the size range that extends down to 1 metre. Below that size, one could say that there are no asteroids, because they are then called meteoroids. Out of the nearly 1 million known asteroids, 500,000 have precisely determined orbits. Most lie in what is known as the 'asteroid belt' between Mars and Jupiter (21–33 units away from the Sun, as measured by Bode's Law). The first asteroids to be discovered were the larger ones. Ceres is the largest at 950 kilometres in diameter. It is regarded as a 'dwarf planet' and looks like Mercury or the Moon. The next largest is Vesta, just over 500 kilometres in size, with a large hole in its south pole area. A meteor impact on Vesta in the past ejected a number of fragments which became a family of small, co-orbiting asteroids and meteoroids.

Like these small bits from Vesta, many asteroids are irregular fragments from the collision of one asteroid with another. Most asteroids are potato shaped, and thought to be planetesimals (small proto-planets); most planetesimals gravitated into larger masses and formed the major planets, but the strong gravity of Jupiter affected those planetesimals in the asteroid belt and prevented them from settling together and congealing into a single planet.

Many asteroids have been kicked out of the asteroid belt by near-miss collisions and encounters with large planets, some of them having been ejected into interstellar space. Some fall in towards the Sun and, if they come near the Earth, they become a natural hazard both for astronauts and for life on the surface of our planet. A few small asteroids have impacted the Earth and burnt up in the atmosphere. Some asteroids pass close enough to the Earth that they can be imaged by radar, although none of these has come close enough to pose a danger to the Earth's inhabitants.

A few asteroids have been visited and imaged at close range by spacecraft, in recognition of their scientific importance as remnants of the early Solar System. On its way to Jupiter, the Galileo spacecraft visited Gaspra in October 1992 and Ida in August 1993. It discovered that Ida had a satellite, which was later named Dactyl – the first asteroid satellite discovered. Other spacecraft have been programmed to fly close to asteroids to take a good look in pursuit of their main missions: Deep Space 1 visited asteroid Braille in 1999; Stardust visited Annefrank in 2002; Rosetta visited Šteins in 2008 and Lutetia in 2010; China's Chang'e 2 flew past Toutatis.

The first mission dedicated to an individual asteroid was the Near Earth Asteroid Rendezvous (NEAR) mission, which targeted Mathilde in June 1997, and then Eros, which the spacecraft orbited for a year until it was skilfully landed onto Eros's surface on 14 February 2001, the touch down interrupting the last picture frame being transmitted back to Earth. The surface of Eros is rubble-strewn with boulders that come mostly from the meteor impact that caused its largest crater. In 2005, the Japanese Hayabusa probe studied asteroid Itokawa, and returned samples of its surface to Earth. NASA's Dawn spacecraft inspected Vesta for a year starting in July 2011 and then went on to Ceres, entering into orbit around it in 2015. Finally, in 2019 the Japanese Hayabusa2 probe visited Ryugu, landing probes on it to gather and return samples to Earth. Returned samples of asteroids are important because they could be the same as the material from which the Solar System formed.

Pluto

A planet deliberately sought – but not a planet,
and discovered by accident

> Then felt I like some watcher of the skies
> When a new planet swims into his ken;
> Or like stout Cortez when with eagle
> eyes he star'd at the Pacific.

John Keats, 'On First Looking into Chapman's Homer', 1817

The discovery of Uranus, the asteroids and Neptune made scientists suspect that there might be even more planets in the Solar System. This suggestion was reinforced by the fact that, by the end of the nineteenth century, it was apparent that Neptune was drifting off its calculated orbit. One speculation, offered by several scientists, was that a large planet could be hiding in the darkest and most distant reaches of the Solar System.

Mindful of Le Verrier's and Adams's discovery of Neptune as the source of a gravitational attraction that pulled Uranus off course, the American astronomer William Pickering calculated where the undiscovered planet might be. Pickering unsuccessfully searched the photographic archive of the Mount Wilson Observatory, California, for images that might show the mysterious new planet. Explaining his lack of success, in 1911 the Indian astronomer Venkatesh Ketakar published a calculation that posited details of the orbits of *two* hypothetical planets beyond Neptune, a model based on the assumption that the gravity generated by each of the three bodies affected the orbit of the others.

The American businessman Percival Lowell took a more pragmatic approach. Lowell was a member of a wealthy, influential Boston family, who had trained at Harvard as a mathematician.

From the age of thirty-eight he devoted himself to astronomy, moving in 1894 to Flagstaff, Arizona, where the clear skies were favourable for astronomical observation. He built a private observatory to study the planet Mars and, starting in 1906, to search for the trans-Neptunian planet, which he termed Planet X. He devoted ten years to his search, concentrating on the regions indicated by Pickering.

How would you find a distant planet? The stars are fixed in position relative to each other, but planets are in orbit and change position quickly. Lowell and his assistants repeatedly photographed the sky in the search regions to check for star-like images that moved. Using this method, Lowell discovered 515 asteroids, but no Planet X.

Lowell Observatory remained in operation after Lowell's death in 1916 and recommenced the search for Planet X in 1927 under its new director, Vesto Melvin Slipher. In December 1929, he hired an amateur astronomer, Clyde Tombaugh, as an assistant to take pairs of photographs two weeks apart. The pairs were put side by side in a viewing device called a 'blink comparator'. Its operator rapidly shifts a mirror back and forth to view each photograph alternately. Any image that has moved between the exposures leaps from one position to another, and is readily identifiable. At first Slipher and his brother worked the comparator, but Tombaugh produced pictures faster than the Sliphers could process them and the brothers became bored. They delegated the task of inspecting the pictures to Tombaugh, who discovered Planet X in February 1930, at the age of twenty-four.

Tombaugh and the Sliphers received lots of advice on the name of the new planet. Lowell's widow suggested that the planet should be named after herself. The astronomers at the Lowell Observatory took a dim view of this, given that she had been trying for ten years to get her hands on Lowell's endowment for the observatory. The name Pluto was proposed by Venetia Burney, an eleven-year-old schoolgirl living in Oxford who was interested in classical mythology, and suggested the name of the Roman god of the underworld because Planet X was presumably dark and cold. She made the suggestion to her grandfather, a librarian at the university, after he read a newspaper article to her about the planet that raised the issue of its name. He passed the suggestion to the Oxford astronomer Herbert Hall Turner, who passed it on to his American colleagues. The name found unanimous favour in a vote of the Lowell Observatory astronomers and was announced in May 1930. A strong point in its favour was that the name started with Percival Lowell's initials, and neatly got around the historical disinclination of astronomers to name planets after people. It left the way open, however, for Pickering to aggrandize himself with the claim that PL stood for Pickering-Lowell.

There are pre-discovery photographs of Pluto dating back to 1914, including two images taken before Lowell's death. Pluto was much fainter than had been expected and was therefore overlooked in these early pictures. Ironically, it seems that Pluto is not massive enough to have caused the deviation of Neptune from its predicted orbit – the discrepancies were the result of errors in estimating the masses of the other planets, and using modern, more accurate values it has been shown that there is no actual deviation in Neptune's orbit. Thus, the reason that Tombaugh discovered Pluto was not that he searched where the planet was calculated to be, but that he diligently searched at all.

In powerful telescopes Pluto shows an unusually large and strangely broadened image, and it looks big – but it is a lot smaller than it looks. Astronomer Jim Christy of the US Naval Observatory in Washington, DC, discovered the reason for Pluto's odd appearance in 1978, noticing that in especially clear photographs of Pluto taken at regular intervals, the elongation seemed to rotate around Pluto. This turned out to be because Pluto has a moon that is nearly as large and as bright as the planet itself (the moon is half of Pluto's diameter). The moon is distinctly separate from Pluto but still unusually close for a satellite, at a distance of only seventeen times its radius. The moon was named Charon, after the mythological ferryman to the Underworld, Hades, Pluto being the Roman lord of that domain.

The existence of Charon was confirmed between 1985 and 1990 when the orbital plane of Pluto and Charon became visible edge-on from Earth, producing the expected series of mutual eclipses as Pluto and Charon passed in front of each other in turn. Since the right alignment occurs for only two 5-year intervals in Pluto's 248-year orbit, it was lucky that this happened so soon after Charon's discovery. Any lingering doubts about the existence of the moon were removed when the Hubble Space Telescope (HST) imaged Pluto and Charon side by side in 1994. Between 2005 and 2012 the HST discovered a further four, smaller moons orbiting around Pluto. They were given names that are appropriately connected to the domain of Pluto: Styx, Nix, Kerberos and Hydra.

It is likely that Pluto and Charon formed as separate objects in a near-circular, low-inclination orbit beyond Neptune. Neptune perturbed their orbits and they got fed into the same eccentric, steeply inclined orbit, controlled by Neptune. Eventually Pluto and Charon collided and formed the present binary system.

Pluto is an unusual planet; in fact so unusual that in 2006 it was removed from the status of 'planet' by the International Astronomical Union and termed a 'dwarf planet'. It is a member of the Kuiper Belt. Pluto has been visited once, in a fly-by in 2015 by the New Horizons spacecraft. It is covered with mountains of ice and smooth, flat plains of frozen nitrogen. One especially remarkable plain called Sputnik Planitia appears to be a filled-up meteor crater. The meteor impact melted frozen nitrogen and water-ice under Pluto's surface, the liquidized mixture oozing up and filling the crater to the brim.

The Kuiper Belt

The frontier of the Solar System

> The most important thing is not to work on things that other
> people are working on because otherwise all you'll do is get
> the same result as everybody else and you won't make any
> discoveries, you'll just confirm what's already known.

David Jewitt, *On Asteroids – The Good, the Bad and the Ugly*, 2014

Following the discoveries of Neptune and Pluto, astronomers began
to suspect that the Solar System was much larger than previously
thought, and that its frontier extended well beyond the orbits of
the known planets. This theory, which originated in the 1950s, was
proved forty years later with the discovery of the Kuiper Belt, a
belt of asteroids and comets that lies beyond Neptune and is made
up of material left over from the formation of the Solar System.
Many of the comets that we see from Earth are objects that have
travelled from the Kuiper Belt.

Like Pluto and Neptune, the Kuiper Belt was 'discovered' in
theory long before it was actually seen. American professional
astronomer Gerard Kuiper (rhymes with 'viper') is conventionally
credited with proposing the existence of the Belt, which he put
forward at a symposium in 1951, hence the Belt's name. However,
another American astronomer, Frederick Leonard, had mentioned
the idea in print as early as 1930, in a publication for amateur
astronomers that was largely ignored by professionals.

Irish amateur astronomer Kenneth Edgeworth laid out the idea
more clearly in scientific papers that he wrote in the late 1940s.
Edgeworth was a soldier and engineer who won the Military Cross in
the First World War and had worked in the post and telegraph office
in the Sudan, taking up an interest in astronomy when he retired.

He wrote articles for amateur and professional journals about the origin of the stars and the Solar System. In 1949 he published an account of the formation process of stars, in which he speculated that the formation of the Sun might have left a debris field beyond Pluto. In its first forty years Edgeworth's article was referenced only a few times, but it is much better known today – the 'Kuiper Belt' is therefore sometimes called the 'Edgeworth–Kuiper Belt'.

These astronomers argued that there was no reason why the Solar System should end abruptly at Neptune or Pluto. In its modern form, the basic argument would be that the solar nebula, from which the planets were formed, extended well beyond Neptune, but planets as large as Neptune could not form further out in the Solar System. Because of the low gravity in the outermost reaches of the Solar System, everything in that region moves so slowly that collisions in which smaller bodies stick together to form bigger ones are very infrequent, and therefore only small bodies can accumulate. Even some of the original smaller bodies survive. Because it is so cold in those distant regions, these small bodies would be ice-rich asteroids, or comets. In fact, some of the comets that we see from Earth have come from this population, a possibility first proved theoretically by pioneering mathematical work by American astronomer Paul Joss in 1970 and Spanish astronomer Julio Fernández in 1980.

In 1988 the Canadian astronomers Martin Duncan, Tom Quinn and Scott Tremaine used comprehensive computer simulations to show that when individual bodies in the Kuiper Belt encountered each other in near-misses, their orbits might indeed be disturbed dramatically. Many of the Kuiper Belt objects are flung out of the Solar System into interstellar space but some fall inward towards the Sun. Some of these 'fallen objects' become trapped into short-period orbits by the massive planets that they encounter as they fall towards the Sun. If this happens, the Kuiper objects become comets, repeatedly passing by the intense heat of the Sun and progressively melting until eventually they break up, the residual dust causing meteor showers. As they studied the properties of such short-period comets, Duncan, Quinn and Tremaine concluded that all of these comets had originated from a single flat disc of material, because their orbits around the Sun were confined to a single plane. The plane of the comets' orbits was close to the plane of the Earth's orbit, which suggested that the disc of comet and asteroid material had originated during the formation of the Solar System. This disc is the Kuiper Belt, with Neptune at its inner edge.

In 1987 British-born astronomer David Jewitt was becoming increasingly puzzled by the emptiness of the outer Solar System: 'It was just freaky.' He persuaded Jane Luu, a colleague at the University of Hawaii, to help him find what was beyond Neptune and Pluto, because, as he told her, 'If we don't, nobody will.' They set out on a search using the same technique used by David Tombaugh to find Pluto, repeatedly imaging an area of the sky to find moving objects in the Solar System – though using digital electronic detectors called Charge Coupled Devices (CCDs) instead of photography. In August 1992, after a five-year search, they found the first member of the Kuiper Belt, which was catalogued as '1992 QB1'. This was the archetypical Kuiper Belt Object, and gives its 'QB1' catalogue number to the name of similar objects in the Kuiper Belt, which are collectively called cubewanos.

The small planets or comets in the Kuiper Belt are called Kuiper Belt Objects, or perhaps more neutrally Trans-Neptunian Objects or TNOs. Thousands of TNOs have been discovered, and there are probably over 100,000 of them with diameters larger than 100 kilometres. The cubewanos, which account for most of these, have nearly circular orbits that lie in the same plane as the major planets. They are the original planetesimals and have always orbited this way. But some of the objects have orbits that have been strongly disturbed. A few are in large, eccentric orbits, perhaps put there by a long-ago interaction with Neptune or disturbed by a passing star.

About a quarter of the known TNOs are in resonance with Neptune. This means that the object makes a certain number of orbits around the Sun for every orbit completed by Neptune. Pluto does this, making three orbits for every two that Neptune makes, so it is in the '3 to 2' resonance. A large number of TNOs are also in the same '3 to 2' resonance, and are consequently called Plutinos. Because it orbits together with a large group of TNOs, and is physically distinct from the other planets of the Solar System, Pluto is considered the largest known object in the Kuiper Belt, not a planet. This is the stance that informed the decision of the International Astronomical Union in 2006 to revoke Pluto's status as a planet in 2006.

Pluto was investigated in 2015 by the New Horizons spacecraft, which flew by and on into the Kuiper Belt. The spacecraft's future trajectory was examined by the Hubble Space Telescope, to see what TNOs were within its reach. The target was detected and chosen, and named Ultima Thule, after the mythological island to the north of Britain that became the generic name in the Middle Ages for a distant and unseen land. It orbits with a period of 298 years at an average of 44.5 times the Earth–Sun distance from the Sun. New Horizons flew by Ultima Thule in 2018–19. It proved to be two lumps in contact. Seen from some angles it looks much like a snowman. It is thought that two TNOs collided gently and fused together. It has few craters, presumably because potential colliding objects are so scarce at this distance from the Sun.

Ultima Thule is the most distant object in the Solar System that has been visited by a spacecraft. Even further from the Sun are three spacecraft (Voyagers 1 and 2 and Pioneer 10), some comets that have very eccentric orbits, and further TNOs, one of them at a distance of 120 times the distance of the Earth from the Sun. The largest is Eris, a TNO as large as Pluto (2,400 kilometres in diameter) and has a satellite, Dysnomia. It has a very eccentric orbit with a period of 557 years. As of 2018, it lies 96 times further from the Sun than the Earth. This is the frontier of our Solar System.

Meteors and Meteorites
The sky is falling!

> All you that do behold my stone
> O: think how quickly I was gone:
> Death does not always warning give
> Therefore be careful how you live.

John Shipley's headstone in Wold Newton churchyard, 1829

Meteors are what are popularly called 'shooting stars'. An interplanetary piece of rock or dust, called a meteoroid, may fall onto the Earth if its orbit crosses the Earth's orbit. As the meteoroid enters the Earth's atmosphere, its surface melts because of friction with the air; its outer layers vaporize, and the gases glow. Small meteoroids completely disintegrate in the atmosphere. If the meteoroid is large or robust, it can survive the fall and reach the ground as a rock or iron lump, in which case it is called a meteorite. It may hit the ground so hard that it makes a crater.

Meteors are relatively common, but meteorites are much rarer. For centuries, people have recorded unusual stones that fell from the sky, sometimes regarding them as sacred. Some meteorites have been seen to fall, accompanied by the sound and flash of an atmospheric explosion. In 1492 a huge triangular stone noisily made a metre-deep hole in a wheat field outside the small town of Ensisheim, Alsace, witnessed by a young boy. Because the Emperor Maximilian decided that the fall was a good omen, the 'Thunderstone of Ensisheim' is preserved in the Regency Palace there. One of the best-documented early falls occurred near the village of Wold Newton near Scarborough in England, where in 1795 a 17-year-old ploughman, John Shipley, saw and heard a 25-kilogram meteorite impact on the ground 8 metres away from him, and was showered by earth from the resulting 50-centimetre-deep crater. This meteorite is now in London's Natural History Museum, and a pillar marks the spot where it fell. In 1992 a meteorite damaged a Chevrolet Malibu car in Peekskill, New York, which was later put up for sale at a premium price – one of the few cases where a dent in a car's bodywork elevated its value.

The fourth-century BCE Greek philosopher Aristotle explained meteors as a wholly atmospheric phenomenon. He thought that everything was composed of different proportions of four basic elements: 'earth', 'water', 'air' and 'fire', the latter being something like 'inflammable material'. In his book *Meteorologica* Aristotle suggested that thin streams of a mixture of 'fire' and 'air' rose to the top of the atmosphere. Ignited by the rotating motion of heavenly bodies turning around the stationary Earth, the exhalations burst into a flame, like sparks off a grinding machine, making a 'shooting star'. This point of view about the origin of meteors is the reason why they have a name that seems to connect them with weather. Variations of the explanation persisted until the end of the eighteenth century CE, though Aristotle considered meteorites unrelated phenomena caused by bits of volcanoes that had been launched into the sky by distant explosions. Aristotle's view was that stones could not originate from the sky, and for over two millennia, learned people dismissed accounts of meteor strike as peasants' fables. However, at the beginning of the nineteenth century this view was suddenly replaced by a new paradigm in the face of the overwhelming scientific evidence.

In 1794 the German physicist Ernst Chladni laid out the connection between meteors and meteorites. He studied a 700-kilogram iron meteorite that had been found in Siberia. Its surface was blackened and had been melted. Its composition was similar to other meteorites that had been found in widely distributed areas around the world. These meteorites were mainly iron, but had been found in places that had no iron deposits. Chladni concluded that meteorites must have the same source and that this source must cover the whole Earth – they had fallen from space.

Chladni's hypothesis was confirmed by the French scholar Jean-Baptiste Biot, who was sent by the French Academy of Sciences to investigate reports of many stones falling from the sky at L'Aigle, Basse-Normandie, in 1803. Biot had firmly believed in the Aristotelian explanation for meteorites, but two pieces of evidence changed his mind. One was the number of reports by respected people who had actually witnessed the 'fall of a rain of stones thrown by the meteor'. The other was the sudden appearance across the area of stones that had no similarity with any kind of mineral or human artefact from the region.

The stone in the Temple of Apollo at Delphi, Greece, known to the ancients as the *omphalos* or 'navel of the world', was a meteorite said to have been thrown to earth by the god Cronus when he created the Universe. In the Great Mosque of Mecca, the *hadschar al aswad* is a sacred 'Black Stone' kept in the Kaaba, the axis of the Islamic world. Although the stone has never been examined scientifically, it is thought to be a meteorite, said to have been given to Abraham by the archangel Gabriel and at one time possessed by the prophet Mohammed. The 14-tonne Willamette iron meteorite, now on display in the American Museum of Natural History in New York, was originally used by the tribes of the native Clackamas people in Oregon in pre-hunting rituals to harden their weapons.

The best place to discover meteorites is Antarctica. They are relatively easy to find on the white surface of ice, with the nearest terrestrial rock 3,000 feet underneath. Moreover, the flow of ice down valleys under the snow cover concentrates the meteorites into particular places. The first Antarctic meteorite was discovered in 1912, by a member of Douglas Mawson's Australian expedition. In 1969 Japanese glaciologists discovered nine meteorites within 3 kilometres of each other – the meteorites were of five different types and therefore not fragments from the same fall. This find emphasized the importance of Antarctica as a place to discover meteorites. The Japanese National Institute of Polar Research and the University of Pittsburgh set up expeditions to Antarctica in the mid-1970s, which led to the establishment of the Japanese Antarctic Meteorite Research Center in Tokyo and the US Antarctic Search for Meteorites programme (ANSMET), now led by Scott Sandford. Tens of thousands of meteorites have since been collected from the continent.

Meteorites are valuable sources of information about the makeup of other planets in the Solar System. Roberta (Robbie) Score of ANSMET discovered ALH84001 (its number signifies that it was found in the Alan Hills icefield in North Victoria Land in 1984), which originated from the planet Mars, as shown by a comparison of the gases trapped in it with measurements of the Martian atmosphere made by the Viking lander spacecraft. ALH84001 is one of about two hundred meteorites that have been found that were ejected from the surface of Mars by the impact of a comet or asteroid. They often contain minerals or molecules that do not naturally occur on Earth. It is the hope that they will one day offer clear evidence that there is, or has been, life on Mars. Other meteorites come from the Moon and from the asteroid Vesta.

Meteor Showers
'In the middle of the night, stars fell like rain'

> In the gloomiest period of the war, [Abraham Lincoln] had
> a call from a large delegation of bank presidents. In the talk
> after business was settled, one of the big dons asked Mr.
> Lincoln if his confidence in the permanency of the Union was
> not beginning to be shaken—whereupon the homely President
> told a little story: 'When I was a young man in Illinois, I
> boarded for a time with a deacon of the Presbyterian church.
> One night [in 1833] I was roused from my sleep by a rap at
> the door, and I heard the deacon's voice exclaiming "Arise,
> Abraham, the Day of Judgement has come!" I sprang from
> my bed and rushed to the window, and saw the stars falling
> in great showers! But looking back of them in the heavens I
> saw all the grand old constellations with which I was so well
> acquainted, fixed and true in their places. Gentlemen, the
> world did not come to an end then, nor will the Union now.'

Walt Whitman, *Prose Works*, III. *Notes Left Over*, 17.
A Lincoln Reminiscence, 1892

On any night it is possible to see sporadic meteors, but regularly, on certain days of the year, meteors come in showers. The first record of a meteor shower dates from 16 March 687 BCE, when astronomers of the Chinese Chou dynasty noted that: 'In the middle of the night, stars fell like rain.' Mistaken for the apocalyptic collapse of the heavens in ancient times, and for V-2 rockets during the Second World War, meteor showers are caused by clouds of dust and rock that have regular orbits and are closely associated with comets.

The meteors of a shower have a common origin in a single comet (or in one known case, an asteroid). Meteoroids released from the parent body spread along its orbit and form a meteoroid stream, which might intersect the Earth's orbit. When the Earth passes through the stream, lots of the meteoroids shower into the Earth's atmosphere. These meteoroids travel in parallel. Seen in perspective from the Earth's surface, the meteors appear to radiate from the same point, just as parallel railway tracks do. The radiant (that is, the vanishing point of the paths of the individual meteors, observed from Earth) lies in a given constellation or near a given star.

There might be half a dozen sporadic meteors per hour on a normal night, but anything up to tens of thousands of meteors per hour during a shower. The number during an annual shower varies from year to year because the meteoroids travel around their orbit in clumps, with the main clump closely associated with the parent comet or asteroid, and in a given year the Earth might or might not pass through a clump. Also, the meteors' orbit may move or split into sub-streams, so the Earth might pass closer or further from the centre of the stream or between two.

In mid-November of each year occurs the Leonid meteor shower (known by this name because its radiant is in the constellation Leo). The Leonid meteor shower of November 1833 was particularly spectacular, peaking at 1,000 meteors per minute. It was best seen from North America, and was recorded by Native Americans; the calendars ('winter counts') kept by the Sioux tribes name each year after a notable event, and 1833–34 was called 'stars all falling down year', adding: 'They feared the Great Spirit had lost control over his creation.' The shower was referred to by Abraham Lincoln in an anecdote recorded by Walt Whitman, and it inspired the jazz standard 'Stars fell on Alabama'.

This Leonid shower of 1833 was analysed by Denison Olmsted and Catlin Twining of what became Yale University, who discovered the radiant of the shower and realized that the radiant was actually the orbital path of the meteor stream. Later, Hubert Newton, also of Yale, calculated the orbital period of the meteoroids at thirty-three years and identified appearances of the shower dating back to 902 CE.

After Tempel's Comet was discovered in 1866, it became clear that it was the parent body of the Leonids, because its orbit was identical with the orbit of the meteors.

In 1836 a Belgian astronomer and statistician, Adolphe Quetelet, discovered a second shower, the Perseids, which occur in mid-August each year. The 1834 shower had been seen by John Locke, the headmaster of a girls' school, who published a letter about it in the Cincinnati *Daily Gazette*. His account, which also mentioned that he had discovered the shower's radiant in Perseus, went unnoticed by astronomers, most of whom would not have read this local newspaper. Quetelet predicted a display of meteors in August 1837. Edward Herrick, a bookseller–librarian in New Haven, Connecticut, observed the shower on 9 August and identified seven occasions in the past millennium when August meteors had been seen before, ranging from 1029 in Egypt to 1833 in England. He found a reference to a European superstition that the 'burning tears' of St Lawrence are seen in the sky on the night of the 10th of August, this day being the anniversary of the saint's martyrdom in Rome in 258 CE. Evidently the Perseid meteor shower had been known for centuries. In 1867 Italian astronomer Giovanni Schiaparelli discovered that the Perseid meteors came from a stream whose orbit was the same as the bright comet of 1862, Comet Swift–Tuttle, named after the two American astronomers who had discovered it.

Meteor showers can only be seen at night (unless the meteor is exceptionally bright). If a shower occurs at a time of the year when the Sun lies near to the direction of the radiant then the shower remains invisible, because you can't see meteors during daytime. However, in 1944 radio engineer James Hey discovered that meteors generate radar echoes, which can be detected by day as well as by night. In September of that year, during the Second World War, V-2 missiles were being launched on London from Germany. As a member of the British Army Operational Research Group, Hey developed a radar system to detect incoming missiles in the hope that the civilian population could be given warning of an attack to minimize casualties. Hey's system could indeed detect incoming V-2 rockets but also gave many false alarms, detecting launches of rockets when none was reported by spies and predicting attacks that never materialized.

After the war had finished, Hey set out to discover what had caused the false echoes. He discovered that his spurious radar echoes occurred when a meteor trail passed through the radar beam. The radar reflection was from the ionized air produced in the meteor trail. Final proof came from an organized campaign, when Hey and colleagues at other stations coordinated to look for radar echoes during the Giacobinid meteor shower of October 1946, a shower associated with Comet Giacobini–Zinner. They saw ten thousand radar echoes per hour rather than the usual two or three. Using radar echoes, Hey and his team quickly discovered several new daytime meteor showers: the Arietids, the Zeta Perseids and the Beta Taurids.

The Earth's Magnetosphere
Our defence against the Sun

And now the Northern Lights begin to burn, faintly at first, like sunbeams playing in the waters of the blue sea. Then a soft crimson glow tinges the heavens. There is a blush on the cheek of night. The colours come and go; and change from crimson to gold, from gold to crimson. The snow is stained with rosy light. Twofold from the zenith, east and west, flames a fiery sword; and a broad band passes athwart the heavens, like a summer sunset. Soft purple clouds come sailing over the sky, and through their vapoury folds the winking stars shine white as silver.

Henry Wadsworth Longfellow, *Driftwood, Frithyol's Saga*, 1837

The Earth is a giant magnet. Its liquid-iron core generates a magnetic field around the planet that extends outward as far as the Moon. This magnetic field not only causes magnetized compass needles to point north, but also shields the Earth from lethal doses of solar radiation. Without its magnetic field, the surface of the Earth would resemble the desolate landscape of Mars.

The Earth's magnetic field is caused by the circulatory motions of the liquid-iron core of the Earth, which generates currents and a magnetic field much in the same way that dynamos and generators do. The region that is subject to the magnetic field's direct influence is called the magnetosphere (the term was coined by Cornell University scientist Tommy Gold in 1959) and extends into space towards the Moon. The magnetosphere was the first major scientific discovery of the space age.

For centuries, sailors in many cultures knew of the lodestone, which indicated the direction north. If freely suspended – for example, floating on cork on the surface of water – some minerals, like magnetite, constitute a magnetic compass, by which it is possible to navigate if the shore or the stars are not visible. The magnetic properties of the lodestone can be transferred to an iron needle for greater clarity of direction.

In 1576 a ship's instrument maker, Robert Norman, noticed that in London a magnetized needle not only turned to point north, but also tended to dip down below the horizontal by about 70 degrees. In 1600 the English physicist William Gilbert realized that this was because the needle was following lines of magnetic force that converged down towards the surface of the Earth, and that the Earth is itself a magnet whose effects extend out above the surface and must continue into space. The Earth has two concentrations of the lines of force at its magnetic poles, where, as Gilbert realized, a magnetic compass needle would point vertically downwards.

Between 1698 and 1700 the astronomer Edmond Halley combined his own magnetic survey of the Atlantic Ocean with other people's measurements, producing the first map of the world showing the direction in which a magnet pointed at any given location. Magnets always point a little off the true north because the north magnetic pole is not identical with the North Pole of the axis of rotation of the Earth. There were repeated surveys in the centuries since then, but in 1957–58 there was a major, coordinated global effort to study geomagnetism, called the International Geophysical Year (IGY). Its scientific role was usurped when it became the public arena for the military objectives of the Cold War between the USA and the USSR.

American scientist James van Allen became a key figure in the execution of the IGY. He had worked on high-altitude experiments, using V-2 rockets at first. He had then developed small so-called 'sounding rockets' and a hybrid vehicle called a rockoon, a rocket taken to altitude and launched from a balloon. He used all these to 'sound' the upper atmosphere.

On 4 October 1957 the USSR reached beyond the confines of the Earth and launched Sputnik 1, the first artificial satellite to orbit the Earth. This event is reckoned as the start of the 'Space Race' between the USSR and the USA. Early in 1958 the USA responded to Sputnik by launching its first space probes, Explorer 1 and Explorer 3, both carrying experimental equipment with which van Allen measured the density of charged particles in space.

Charged particles from the Sun and from outer space are caught by the Earth's magnetic field and funnelled through the magnetosphere into the magnetic polar regions, as discovered by the Norwegian physicist Kristian Birkeland in 1895. Birkeland put a magnetized iron sphere – a terrella, or 'little Earth' – in a vacuum chamber and aimed a beam of electrons towards it. He saw that the electrons were steered by the magnetic field to the terrella's magnetic poles. This experiment replicated in miniature the phenomenon of the aurora, or Northern and Southern Lights. When electrons strike the Earth's atmosphere, they produce a colourful, shimmering aurora that can be readily seen at polar latitudes. Birkeland's terrella was an experimental way to simulate the magnetosphere, which is nowadays carried out by numerical simulations by computers.

Van Allen discovered that the Earth was encircled by a doughnut-shaped region of charged particle radiation. This radiation was trapped within 'magnetic bottles' that were enclosed by the Earth's magnetosphere. The doughnut-shaped region of radiation was named the 'Van Allen Belt' in honour of its discoverer. A second, outer belt was identified in 1958 by the Pioneer 3 lunar probe (which failed to get to the Moon but nevertheless reached an altitude of 101,000 kilometres), and also detected by the Sputnik 2 and 3 satellites. The Van Allen Belts were the first major scientific discovery to be made as a result of space exploration.

The Van Allen Belts reach from the top of the Earth's atmosphere out to a distance of about 3.5 times the diameter of the Earth. They are major components of a system of electric currents and high-radiation regions that have been mapped by a succession of space satellites. The satellites sent into the high-radiation regions have to be especially robust because the particles deleteriously affect the environmental conditions for spaceflight. The particles also affect the transmission of power and electrical signals at the Earth's surface. All these effects are known as 'space weather'. They are caused by storms produced by the Sun and are associated with sunspots.

The first signs that the Sun causes changes in the magnetosphere were noticed in the nineteenth century. In 1843, Heinrich Schwabe, a German apothecary and amateur astronomer, discovered that sunspots came and went on a ten-year cycle (now more accurately reckoned at eleven years). The explorer Alexander von Humboldt popularized Schwabe's work. Consequently, when an army officer named Edward Sabine discovered that magnetic storms (violent vibrations of a compass needle) were more frequent at intervals of ten years, he suggested that this was related to the solar cycle. Sabine's theory was dramatically confirmed when English amateur astronomer Richard Carrington discovered a large flare on the Sun in 1859, which was immediately followed by a violent magnetic storm and aurora.

Solar wind – the second major discovery of the space programme – is the actual mechanism in the Sun that affects the Earth's magnetosphere. This is a constantly flowing but erratic stream of charged particles that emanates from the Sun and impinges on the Earth – or rather onto the Earth's magnetic field. The behaviour of comet tails – which always point away from the Sun, regardless of the direction in which the comet is travelling – gave early indications that something was flowing outwards to sweep the tails away, like the loose ends of a scarf blowing in the wind. Physicists Eugene Parker, Ludwig Biermann and Hannes Alfvén independently came to the conclusion in the 1950s that there existed a relentless outward flow of charged particles from the Sun: the solar wind. The solar wind was confirmed by the Soviet lunar probes Luna 1, 2 and 3 as they transited from the Earth to the Moon in 1959 and by the US Mariner 2 probe as it travelled to Venus in 1962. Except at the poles, the magnetic field of the Earth defends the Earth's atmosphere and surface from the solar wind. If the magnetosphere did not exist, the Earth's atmosphere would be blown off by the solar wind and its surface would be exposed to lethal radiation. This seems to be what happened to the planet Mars when its magnetic field died away to almost nothing. By contrast, the Earth's magnetosphere is unusually large because the iron core of the Earth is unusually large, being the amalgamation of two iron cores in the collision that formed the Moon.

The magnetic field of the Earth is always shifting, because it is sustained by the circulatory motions in the Earth's liquid-iron core, which are erratic and oscillate. This causes the magnetic field of the Earth to drift and tilt, as English astronomer Henry Gellibrand discovered in 1635. As a result, the magnetic poles wander. The North Magnetic Pole has been moving quickly northwards from Hudson's Bay since 1990 and lies in the Arctic Ocean, near the North Pole. In 2020 it is expected to cross from the sea north of Canada and Alaska into the sea north of Siberia. The South Magnetic Pole is in the Ross Sea just off the continent of Antarctica, towards Australia. The Earth's magnetic field also changes strength, and even reverses polarity (the North Pole changes to the South Pole and vice versa) every 250,000 years or so. It isn't known how quickly this happens, or whether for some periods of time the Earth's magnetic field becomes so weak that the magnetosphere is turned off. If so, the atmosphere and the surface of the Earth would be exposed to the solar wind. It is not known what happens temporarily to the natural environment at that time.

Comets

Dirty snowballs?

He, first of men, with awful wing pursued
The comet through the long elliptic curve,
As round innumerous worlds he wound his way,
Till, to the forehead of our evening sky
Returned, the blazing wonder glares anew,
And o'er the trembling nations shakes dismay.

James Thomson, 'To the Memory of Sir Isaac Newton', 1727

Although comets are no longer superstitiously associated with bad fortune, the mystery of where they come from and what they are made of continues to intrigue astronomers. Recent images from space probes show that comets are made of ice and fine dust. It is possible that water in our oceans and perhaps even the molecular seeds of early life were brought to Earth by comets.

Comets are small bodies that orbit in the Solar System, like planets. However, unlike planets, whose orbits are nearly circular and confined largely to one plane (the 'ecliptic'), comets' orbits are highly eccentric and may be inclined upward or downward at any angle. Aristotle considered comets to be atmospheric phenomena, but in 1577 Tycho Brahe discovered by measuring the parallax of a comet that it was located beyond the Moon, and astronomical in origin.

Kirch's Comet of 1680 was the first comet discovered by telescope. Isaac Newton discovered that the comet was following a near-parabolic orbit around the Sun and conformed to Kepler's laws. It was the comet that proved Newton's law of gravitation. Newton's colleague Edmond Halley used Newton's laws to discover that the orbits of three comets, which appeared in 1531, 1607 and 1682, were very similar. Halley suggested that all three comets were actually the same object, revisiting every seventy-six years on an elliptical orbit, not a parabolic one, and predicted the next appearance in 1758 or early 1759. This happened after his death, when what is now called Halley's Comet was rediscovered by Johann Georg Palitzch, a farmer from Saxony and an amateur astronomer. During the 1986 reappearance, two Russian Vega spacecraft surveyed Halley's Comet from a distance and two Japanese probes investigated its plasma tail, while the European Space Agency probe, Giotto, passed within 600 kilometres of its nucleus – so close that the spacecraft was damaged, hitting a large particle emitted from the comet.

American astronomer Fred Whipple proposed the 'dirty snow-ball' model for the composition of comets. He suggested that a comet's nucleus is made of dust grains cemented together by ices, such as water, ammonia and methane. The ices sublimate to gas and are released from the comet as it warms on approaching the Sun. The vaporization of the ices lets loose the dust grains, which are dragged from the nucleus by outflowing gas. The comet develops a bright, dusty atmosphere, called the 'coma', and 'tails' of gas and dust. But in the early 1950s, Giotto found that Halley's Comet was not really a 'dirty snowball', more a 'snowy dirtball' – its nucleus was a coal-black, peanut-shaped body, about 15 kilometres long and between 7 and 10 kilometres wide; its structural arrangement was dictated by the physical properties of dust rather than ice.

Dust consolidates on a comet's surface as a crusty skin. The action of sunlight on the compounds of which the dust is made creates a sticky substance that coats everything with a black tar: comets are as black as coal. Among the chemical compounds created in this way are organic molecules like amino acids: perhaps these molecules, delivered to Earth in cometary collisions, are materials that helped life to develop here. The European Space Agency's space probe Rosetta orbited Comet 67P/Churyumov–Gerasimenko for two years between 2014 and 2016. The comet was structured like two snowballs consolidated together, with spectacular cliffs and chasms. It may be the outcome of a soft, low-speed collision between two comets that fused together. Unusual, knobbly humps on the comet's surface indicate it is the result of the accumulation of many smaller comets. In 2014 Rosetta loosened a lander, Philae, onto the comet's surface. Unfortunately, it made a bad landing and fell into shade, so scientific results from its mission were limited.

The tails of comets always point away from the Sun. The dust of a comet tail is pushed away from the comet by solar radiation pressure, as was discovered in 1900–1 by the Swedish chemist Svante Arrhenius and the German physicist Karl Schwarzschild. The dust trails of some comets (Halley's Comet, for one) are associated with meteor showers. In 2004 NASA's Stardust probe flew through the tail of Comet Wild 2 and retrieved some of its dust. Some of it was crystalline, 'born in fire' – presumably in the hot inner parts of the nebula that formed the Solar System. It was similar to the material that makes up some asteroids.

In 2005 the Deep Impact probe was crashed onto Comet Tempel 1 and excavated a crater to expose its interior. Most of the comet's ice lies below the surface in accumulations that feed jets of vaporized water that spurt out from the comet. Giotto witnessed how jets on Halley's Comet threw out 3 tonnes of comet material per second. Rosetta witnessed jets being shot from Comet 67P. The gas includes vaporized tarry substances, gases of carbon compounds that make comets some of the smelliest places in the Solar System. Most of the ejected grains of dust are no larger than specks of cigarette smoke. The largest grain detected by Giotto was 40 milligrams, but the large particle that damaged the spacecraft was perhaps as heavy as 1 gram.

Comets also have a second, faint, straight tail of gas, which is pushed back by the solar wind. In 1985, NASA's International Cometary Explorer (ICE) probe explored the gas tail of Comet 21P/Giacobini–Zinner. It is made of plasma, gas that has been ionized by the collision of solar particles.

Comets lose material in their coma and tail at every passage past the Sun, and refreeze once they have left the Sun behind. After a number of passages there is no more loose material and the crusty surface is thick, keeping the comet's icy material trapped below: the comet becomes inactive or extinct. It becomes an asteroid, of the sort known as a 'Centaur', which has signs of fading cometary activity – a very faint tail, an intermittent coma.

It is thought that comets formed when the Sun formed, 4.5 billion years ago, from interstellar ices condensing onto grains of interstellar dust. They were originally planetesimals that congealed from scraps of dust and gas in the pre-solar nebula. Since then, they remained almost unaltered in two cold, outer regions of the Solar System, until they fell towards the Sun, ultimately doomed to melt like snowmen when the Sun rises. Short-period comets come from the Kuiper Belt, which is located in the outer Solar System beyond the orbit of Neptune. The source of the long-period or sporadic comets is thought to be the Oort Cloud, a spherical reservoir of comets surrounding the Solar System. Dutch astronomer Jan Oort discovered in 1950 that many long-period comets fall towards the Sun from a distance of between 20,000 and 200,000 times the distance from the Sun to Earth. Comets that formed inside Neptune's orbit were ejected into distant orbits during encounters with giant planets, and formed the Oort Cloud and the Kuiper Belt. Occasional encounters with each other, or with passing stars or giant clouds of interstellar material, reinject some comets from the Oort Cloud and Kuiper Belt back into the inner Solar System.

It is likely that, early in the history of the Solar System, there were frequent collisions of comets with Earth. Some of our ocean water may have been brought to Earth by comets, although, when Rosetta measured some of the features of the composition of water on Comet 67P, it proved to be dissimilar to the water in Earth's oceans. In addition to water, complex organic molecules (and especially 'prebiotic' organic molecules) could also have acted as seeds for the development of life on Earth.

Astronomical Cycles
The Earth's climate, seasons and the weather

> Nature is an endless combination and repetition
> of a very few laws. She hums the old well-known
> air through innumerable variations.

Ralph Waldo Emerson, 'History', in *Essays*, 1841

As the Sun warms the atmosphere, the land and the oceans, it sets in motion the cycles that generate the Earth's weather. But what triggers the catastrophic chill of an Ice Age? A Serbian mathematics teacher, Milutin Milankovič, suspected that Ice Ages were caused by cyclical changes in the Earth's orbit. Deep below the ocean floor and the Antarctic ice, the proof of Milankovič's theory was waiting to be uncovered.

The Sun warms the Earth most directly in the regions around the Equator. There, sunlight radiates through the atmosphere to warm the ground and surface of the sea. As they warm, the layer of air at the base of the atmosphere becomes less dense and it rises (a process called 'convection'). Cold air is drawn in underneath the rising warm air. At high altitude, the warm air cools and flows away from the Equator, eventually beginning to sink again at about latitude 30° and returning to the equator at ground level. In 1686, Edmond Halley discovered that this cycle was the engine for the major wind systems of the world.

Although this convection cycle creates the wind pattern, the wind does not blow just north and south from the equator. In the days of sailing ships, sea travel was heavily dependent on the 'trade winds', which flow strongly and consistently from the east in the areas around the Equator. Why do the trade winds blow at right angles to what you would expect? George Hadley, an English lawyer and amateur meteorologist, explained the phenomenon in 1735; the closed cell of air cycling from the Equator to latitude 30° and back is consequently called a 'Hadley cell'. Hadley realized the important effect of the rotation of the Earth on wind motion. As air blowing over the Earth's surface moves from a higher latitude region, where the wind's eastward velocity is lower, into a region of higher eastward velocity (at the Equator), the wind picks up a westward motion.

Hadley's intuitive explanation was given a sound physical basis in 1835 by the French physicist Gaspard-Gustave de Coriolis, who applied Newton's fundamental mathematical theories to the problem. Because of the rotation of the Earth, winds deflect to the right in the Northern Hemisphere and to the left in the Southern Hemisphere. This is called the Coriolis Effect. In 1856 Hadley cells and the Coriolis Effect were brought together in a unified theory by the American meteorologist William Ferrel.

The position and the strength of the winds determine the weather, which therefore ultimately depends on the strength of solar radiation. The warmest zone on the Earth is in general the Equator, but more precisely the sub-solar latitude (that is, the area of the Earth that lies directly 'below' the Sun). Because the Earth is tilted at 23.5° to its orbital plane around the Sun, the sub-solar latitude changes position throughout the year, moving from the Tropic of Cancer at 23.5° N in June to the Tropic of Capricorn at 23.5° S in December. This produces the annual cycle of the seasons, which are influenced locally by factors such as latitude, altitude, terrain, vegetation and proximity to oceans, and may be characterized by particular weather phenomena such as a rainy season or a persistent wind like the mistral.

Additionally, the eccentricity of the Earth's orbit around the Sun has a small but noticeable effect on the annual weather cycle. The Earth does not orbit the Sun in a perfect circle, but in a slight ellipse, which causes the distance between the Earth and the Sun to fluctuate at different times of the year. The Earth is closest to the Sun in the first week of January and furthest in the first week of July. This magnifies the effect of the Earth's tilt on the seasons, so in December and January the summer solar radiation is stronger in the Southern Hemisphere than it is in June and July in the Northern Hemisphere, which is why summers tend to be hotter in the Southern Hemisphere.

If the Earth's tilt and its eccentric orbit stayed the same forever, these annual cycles of the seasons would remain the same. However, long-term cycles of change in the Earth's position and orientation cause changes in the seasons over time. The Earth's axis does not always point in the same direction, but precesses in a cone over a period of 26,000 years, which causes the seasons to shift within that timescale. Moreover, the tilt does not remain constant at 23.5°, but nods between 21.5° and 24.5° over a period of 41,000 years. The larger the tilt, the greater the variation of the seasons.

The eccentricity of the Earth's orbit (which produces the hotter summers in the Southern Hemisphere) is also variable, changing between almost 0 and 7%, on a timescale of 100,000 years. At present, the Earth is midway through that cycle, with a difference of 3.4% between January and July.

The total effect of all these orbital cycles on the weather is complicated. Throughout the 1920s and 1930s the Serbian civil engineer and geophysicist Milutin Milanković, in his second career as a mathematics teacher, devoted himself to studying their effects on climate change. For this reason, they are called Milanković (or Milankovitch) cycles. He attributed the Ice Ages (periods when there are extensive ice sheets, like the one in Antarctica) to these cycles, discovering that recent cold periods have occurred approximately every 100,000 years, when all the Earth's different orbital cycles coincided to produce maximum cooling. His discovery was verified after his death by analysis of ocean sediments and Antarctic ice cores, which show isotopic variations in different layers that were caused by temperature differences when the layers were deposited. In the USA, ice cores are stored at the US National Ice Core

Laboratory in a building in Denver, Colorado. It has over 14,000 metres of ice cores from thirty-four drill sites in Greenland, Antarctica, and high mountain glaciers in the western United States. The cores are kept at a temperature of -35 °C, with four levels of backups and safety systems.

These ice cores show that the present Ice Age (defined by glaciologists as a period in which there are extensive ice sheets, like the one in Antarctica) began 40 million years ago. It grew colder during the Pliocene and Pleistocene periods, starting around 3 million years ago, with the spread of ice sheets across the Northern Hemisphere. Since then, glaciers have advanced and retreated every 40,000 to 100,000 years. The most recent retreat of the glaciers ended about 10,000 years ago.

However, Milankovič cycles are not severe enough to alter global temperature by the amounts that are recorded in ocean sediments and ice cores on their own, so there must be other processes, like the greenhouse effect, that amplify the effects of fluctuations in the amount of solar radiation reaching the Earth. Terrestrial volcanism, continental drift and changes in the composition of the Earth's atmosphere all play a part. The recent increase in human-generated carbon dioxide and other greenhouse gases has started to cause changes that are quicker and more extreme than the Milankovič cycles, increasing the Earth's temperature at an unprecedented rate.

I *The Origin of the Milky Way* by Jacopo Tintoretto (*c.* 1575). Jupiter holds his son Hercules, born to the mortal Alcmene, to be nursed by the goddess Juno. Some milk spurts from her breast, forming the Milky Way.

II *William Crabtree Watching the Transit of Venus AD 1639* by Ford Madox Brown (1903). The awestruck draper views the transit of Venus projected into the attic of his shop through darkened shutters and curtained windows, while his wife struggles to keep the children in order.

III The Galilean satellites of Jupiter. Shown to scale in images from the Galileo probe, the four largest moons of Jupiter are, in order of increasing distance from Jupiter (left to right): Io, showing coloured ash drifts from its volcanoes; Europa, showing a crazed pattern of stained ice floes; and Ganymede and Callisto, their patterns of meteor craters pockmarked in the frozen rock.

IV The volcano Maat Mons on Venus. The Magellan spacecraft mapped the lava flows that extend for hundreds of kilometres from the base of the volcano across the fractured plains shown in the foreground.

ABOVE

v Mars. Frosty water-
ice clouds at the poles
and orange dust storms
(lower right) cover the
dark markings of Mars
that lie among the sand
and rock deserts.

LEFT

vi Earthrise as seen
from the Moon. Apollo
astronauts photographed
earthrise from their lunar
orbiter, encapsulating
the growing realization
that our Earth is a planet
with limited resources.

VII Asteroid Bennu. The OSIRIS-REx spacecraft imaged Bennu from a range of 24 km (15 miles) to reveal its boulder-littered surface. The asteroid, which is only 500 metres in diameter, is made of material not much changed from the original composition of the solar system.

VIII Described as the picture of the century, the first close-up photograph of the lunar crater Copernicus was made by Lunar Orbiter 2 in 1966 as it scanned the Moon's surface for possible Apollo landing sites.

ix Comet 67P/Churyumov-Gerasimenko. The Rosetta probe pictured this two-lobed comet as material was being loosened from it to make its tail.

x When Buzz Aldrin roamed around the Apollo 11 landing module, Eagle, in 1969, his footprint demonstrated the engineering qualities of the lunar surface, but also marked the moment and the place at which humans first stepped onto the Moon.

xi An avalanche has fallen from the steep cliff at the edge of the northern ice cap of Mars. The cliff, 700 metres high, is made of layers of water ice mixed with red dust, its upper surface covered by a blanket of frozen carbon dioxide.

XII Comet McNaught. Ripples in the comet's magnificent tail were caused by periodic releases of dust along the curving orbit and the pressure of solar radiation on the dust particles.

XIII Omega Centauri. This, the largest globular star cluster in our galaxy, is such a bright conglomeration of millions of stars that it is visible to the naked eye even though it is more than 15,000 light years away.

XIV Jupiter's Great Red Spot. The 400-year-old storm is an anti-cyclone that rotates counter-clockwise and creates a turbulent zone in the band of clouds in which it lies. Smaller storms come and go nearby, as shown by the Juno spacecraft.

XV Caloris Basin on Mercury is one of the biggest impact craters in the solar system. It triggered volcanoes at its lower rim (visible as orange spots).

Asteroid Impacts
How the Earth developed

> We have learned now that we cannot regard this
> planet as being fenced in and a secure abiding place
> for Man; we can never anticipate the unseen good or
> evil that may come upon us suddenly out of space.

H. G. Wells, *The War of the Worlds*, 1898

A miner's all-consuming but futile quest for riches under a strange
hill in Arizona unearthed treasure of a different kind: evidence of
meteorite impacts on Earth. From a monstrous fireball in Siberia
to mysterious sparkling stones in the wall of a medieval church, the
evidence of meteorite impacts has transformed our understanding
of evolution and geologic change.

US Route 40 runs east from Flagstaff, Arizona, across the dry
Colorado Plateau and passes close to what from a distance looks
like a low hill. This is Coon Butte. It is a complete surprise to
reach the top of the hill and find oneself looking out over a large,
deep, circular crater. The crater is 1,200 metres in diameter and
170 metres deep, and its rim rises 45 metres above the surrounding
plains. Its interior walls have been weathered since it was formed,
and the wall material as well as wind-driven dust from the surrounding plain have raised the floor, so it was originally deeper
and narrower than it is now.

In 1891 the crater was mapped by Grove Karl Gilbert of the US
Geological Survey. In 1895 Gilbert considered the possibility that
the crater had been made by a meteorite impact, but rejected this
explanation on two grounds. First, the amount of ejected material
in the crater walls and on the surrounding plain was no larger than
the hole in the ground. Overestimating how big a meteor would

have to be in order to produce a crater of the size of Coon Butte, Gilbert did not find the large quantity of extra material he expected. Secondly, when he searched the crater floor with a magnetometer for a large mass of iron, Gilbert found nothing: there was no sign that an intact meteorite had buried itself in the ground. Gilbert therefore concluded that the crater was volcanic, like the nearby San Francisco mountains.

Nevertheless, in 1902, on a hotel veranda in Tucson, a Philadelphia mining engineer named Daniel Moreau Barringer heard local gossip from a government agent that the crater was meteoritic. His imagination was fired by the mention of meteors, and the businessman in him was attracted by the possibility that buried under the crater was a large iron-nickel mass, which he could mine and sell – the nickel was particularly valuable. Samples of rocks found nearby on the surface of the plateau contained 5% nickel and traces of iron, mixed with the ejecta from the crater; the iron and nickel fragments were, therefore, evidently coeval with the crater's formation, and Barrington concluded that Coon Butte had been formed by a meteor.

Barringer and a partner, Benjamin Chew Tilghman, bought mining rights to the area containing the crater and began to search for the meteor mass below its centre. The crater was so nearly circular that it seemed logical to assume that the direction of impact had been straight down. By 1908 the pair had drilled twenty-eight holes in the crater floor, but found no meteorite.

Tilghman then noticed that the ejecta littering the surrounding plain was asymmetrical, strewn to the south. Moreover, the southern rim of the crater was raised, as if the meteor had burrowed under it. Barringer experimented by firing projectiles into earth. The craters he produced were always circular, even if the projectile impacted at an angle, but the ejecta from the crater continued the forward momentum of the projectile. On the basis of this experiment, the two men shifted the focus of their search, and drilled a mine shaft into the interior south wall of the crater. Yet they still did not find the mass of iron-nickel. In nearly every hole their drill encountered an isolated hard 'obstruction', which could well have been a fragment of the meteorite, but they were fixated on discovering the motherlode and paid no attention to these small pieces.

Undeterred by the repeated failures of their mining strategy, Barringer and Tilghman presented their theory that Coon Butte was a meteor crater to the Philadelphia Academy of Sciences in 1906 and to geologists at Princeton University in 1909. Barringer's style of argument was belligerent. He heaped scorn on Gilbert, who was a well-respected geologist, and accused established scientists of blind prejudice. This did not endear him to academic audiences and, on most occasions, they could not stop from tittering.

Meanwhile, Barringer's investors were becoming increasingly alienated by his intemperate remarks, and demoralized by the continuing expense of the fruitless search. When the geologist George Merrill theorized that the Coon Butte meteorite would have shattered into small pieces on impact, leaving no single mass of iron-nickel to be found, they gradually withdrew from the project, abuse flung at their departing backs. Disillusioned by Barringer's demand to continue drilling, Tilghman also pulled out.

Barringer found new backers, but by 1928, nervous at the money that they had spent, they consulted the astronomer Forest Ray Moulton at the University of Chicago. Moulton estimated the mass of the Coon Butte meteorite at 300,000 tons compared to Barringer's estimate of 10 million, which reduced the investors' potential return by 97%; the site would be even less profitable if the meteorite had fragmented into a large number of small pieces that would be impractical to collect. In an attempt to reassure them, Barringer sought a second opinion from Princeton astronomer Henry Norris Russell, who confirmed Moulton's calculations. Having spent a fortune of about $1,000,000 (more than ten times that amount at today's value) on his doomed hunt for the meteorite, Barringer died in 1929.

Barringer's discovery that the Coon Butte crater was meteoritic was potentially of greater value than the meteorite itself. Scientists began using Coon Butte as a model to explain similar craters in the Solar System, such as those on the Moon. Barringer's explanation received dramatic support when, in 1908, the region near the Tunguska river in Siberia was rocked by explosions from a monstrous fireball. It was not until 1921 and 1922 that the Russian scientist Leonid Kulik was able to visit the area and found the new impact crater, which had clearly been caused by a meteor and was surrounded by flattened trees. Kulik's work became known in the West by 1928. Nevertheless, the scientific establishment continued to assert that both Coon Butte and the craters on the Moon were volcanic.

It was only between 1957 and 1963 that Barringer's theory about the origin of Coon Butte was decisively confirmed by astrogeologist Gene Shoemaker. He discovered and described the tell-tale geological clues that indicate a crater has been made by a meteorite impact rather than a volcanic explosion. The key sign is the presence of shocked quartz minerals such as coesite and stishovite, which are fused at much higher temperatures and pressures than those generated by volcanic action, and were first identified in craters formed by nuclear weapons testing. Using these criteria, Shoemaker and his wife Caroline identified many of the 160 other impact craters known around the world. One of these was the Ries crater, which surrounds the village of Nördlingen in Bavaria. The walls of St George's church in Nördlingen are made of a rare and beautiful sparkling mineral called suevite, a strongly shocked material generated by the meteorite impact.

Scientists now estimate that Coon Butte crater was formed about 50,000 years ago, when mammoths, sloths, bison and camels roamed the Colorado Plateau. Animals as far as 25 kilometres away would have been killed or injured by the blast. An impact of this size occurs on average only once in a thousand years on Earth. But once every million years or so the Earth is struck by a meteor large enough to devastate a continent. The extinction of the dinosaurs was caused by a meteor impact that, 66 million years ago, created the Chicxulub crater in Mexico, which was discovered in 1978 by oil geophysicist Glen Penfield.

The realization that meteor impacts shape the landscape of the Earth resolved a 250-year dispute among geologists and evolutionary scientists. At the end of the eighteenth century, years before Darwin formulated his theory of evolution, the French palaeontologist Georges Cuvier proposed that individual catastrophes were the major engines of change in the Earth's natural history, causing phenomena such as continental drift and the extinction of species. This theory was widely accepted at the time because it was consistent with creationist biblical interpretations (for instance, those relating to Noah's Flood). The Russian-Jewish psychiatrist Immanuel Velikovsky revived similar views in the 1950s, knitting mythology, archaeology and pseudo-science into fantastic but very popular theories about floods and cosmic fires. Scientists feel instinctively uncomfortable with this worldview, called 'catastrophism', because it suggests that everything on Earth – its ecosystems, weather, geology and geography – is subject to arbitrary events, a 'tale told by an idiot', and therefore cannot be analysed or modelled using scientific methods. Reacting against this approach, geologists James Hutton in the eighteenth century and Charles Lyell in the nineteenth promoted the theories of 'uniformitarianism' and 'gradualism', insisting that geological change occurs slowly over long periods of time, more along the lines of Darwinian evolution. The Scottish geologist Archibald Geikie paraphrased Hutton in 1905, saying 'The present is the key to the past.'

Following Darwin, uniformitarianism became the dominant paradigm of geology for over a century, until Barringer's crater, Kulik's fireball and the Shoemaker's discoveries made their impact, forcing scientists to combine the two opposing views. Cosmic fireballs are not the stuff of legend, but are meteors, which cause dramatic effects on Earth. Cuvier put it well in 1817 in his great work *Of Changes in the Structure of the Earth*: 'Life, therefore, has often been disturbed on this Earth by terrible events – calamities which, at their commencement, have perhaps moved and overturned to a great depth the entire outer crust of the globe…numberless living things have been the victims of these catastrophes….Their races have become extinct.' Scientists now believe that geological history is both uniform and catastrophic, a process of slow and gradual evolution, punctuated by occasional brief transformational events.

The Origin of the Moon
Neither daughter nor sister of the Earth

In all fairness to those who by training are not
prepared to evaluate the fundamental difficulties
of going from one planet to another, or even from
the earth to the Moon, it must be stated that there
is not the slightest possibility of such journeys.

F. R. Moulton, *Consider the Heavens*, 1935

The Moon is not just the Earth's nearest neighbour: it is the passive
guardian of life on Earth, and its dead landscape, visible to the
naked eye, holds the clues to our own planet's origins. When astronauts first walked on the Moon in 1969, they not only disproved
Moulton's prediction, but also brought back dust and rocks that
would reveal lunar secrets.

From time immemorial, people have speculated about the dark
patches on the surface of the Moon: in folklore they are said to
have the shape of a man, a rabbit or a crone carrying firewood. In
the seventeenth century Galileo pointed his telescope at the Moon
and discovered mountains and craters in a barren, dry desert.
More recently, American astronauts and Russian spacecraft have
explored the lunar landscape, and the geology of the rock samples
they collected has shed light on the origins of the Moon. Despite
its peaceful appearance in the night sky, the Moon has a violent
past, having been born in a fiery collision between the Earth and
another planet. Yet without this catastrophic event, the Earth would
not be stable enough to support life.

The Apollo programme of human lunar exploration arose in
response to a challenge issued by President John F. Kennedy at the
height of the Cold War in 1961, as the USA and USSR competed

for primacy in the 'Space Race'. The exploration of the Moon was initially driven by nationalist aims and technological innovations, although a scientific research programme was established to operate in tandem with the landings.

The American programme began with a fleet of twenty-two robot spacecraft launched by NASA to the Moon: the Rangers crash-landed; the Surveyors landed softly; and the Lunar Orbiters, Explorers and early Apollos surveyed the Moon from orbit (plate VI). Later Apollos landed with their human passengers.

Early Moon landings were manned by test pilots and military men with only basic training in geology, and the first landing sites were chosen for feasibility rather than scientific interest. Apollo 11 (plate X) landed on a featureless plain in the Sea of Tranquillity, Apollo 12 in the Ocean of Storms. Later lunar missions targeted areas of geological interest in the hopes of gaining a clearer understanding of the origins and composition of the Moon. Apollos 13 and 14 were meant to land at Fra Mauro, a site made of debris thrown from the meteorite impact that created the massive crater of the Mare Imbrium basin, and would therefore include rocks ejected from deep inside the Moon. The Apollo 13 landing had to be aborted when the spacecraft's service module was damaged by the explosion of an oxygen tank. Apollo 14 was targeted at the same site, landing safely, but the astronauts got lost during the moonwalk, and did not collect the rock samples the geologists wanted. Apollos 15, 16 and 17 were more successful, landing near the Apennine Mountains, the Descartes highlands and the Littrow valley in the Taurus Mountains, sites chosen for their geological significance, with the Taurus-Littrow valley showing signs of recent volcanism. All the Apollo modules that landed on the Moon brought back rock samples for terrestrial analysis – over 300 kilograms altogether. Some small lunar samples from other sites were returned to Earth by three robotic explorers from the Soviet Luna series of spacecraft.

Other rocks from the Moon were found to be made of materials that were melted more recently than the 'Genesis Rock', probably by the impact of a meteor that formed a lunar crater. At the Shorty crater near the Apollo 17 landing site, astronaut Harrison Schmitt, a professional geologist, discovered 'orange soil' containing volcanic glass from an explosive volcanic fire fountain that erupted 3.6 billion years ago.

Before lunar rock samples were available for analysis, theories of how the Moon was made were very disparate, compromised by insoluble riddles, and – in the words of chemist Harold Urey – 'tended to prove that the Moon did not exist'. George Darwin (son of Charles Darwin), Don Wise and John O'Keefe suggested that the Moon was the Earth's 'daughter': that it was formed from material that had spun off from the Earth's mantle. Other planetologists, such as Gerard Kuiper, thought that the Earth and Moon originated as two 'sister' planets during the early formation of the Solar System. Yet others held that the Earth captured the Moon after it had formed somewhere else, either drawing it into orbit intact (Harold Urey), or as a kind of 'Saturn's ring' of pieces that later coalesced into a sphere (Ernst Öpik).

At a conference in Kona, Hawaii, in 1984, a dramatic and radically different theory was adopted to explain the origins of the Moon, incorporating the information from the analysis of lunar rocks. This was the 'collision model', first mentioned in 1946 by Harvard geologist Reginald Daly and revived in 1975 by planetologists William Hartmann and Donald R. Davis and astrophysicist A. G. W. Cameron. They held that the Moon and the Earth both formed from the glancing collision of two planets: the proto-Earth and a planet the size of Mars called Theia (named after the Titan in Greek mythology who gave birth to the moon-goddess Selene). During the collision, Theia and the proto-Earth broke up and reassembled as two new planets, the Earth and the Moon. Both Theia and the proto-Earth had molten iron cores, which coalesced to form the core of one of the new planets – the Earth – surrounded by a thin mantle. Fragments from the mantles of the two colliders coalesced into the other planet, now the Moon. This explains why some lunar rocks had a similar composition to rocks found on Earth, but others did not. The heat of the collision melted the mantle material that formed the Moon, generating KreeP rocks and encouraging volcanic activity. The high heat also vaporized all the water on the Moon, which is why it is so dry. The rapid rotation of the Earth and the tilt of the Moon's orbit were also consequences of the collision.

Without this violent freak event, life as we now know it would not exist on Earth. Compared to other satellites, the Moon is unusually large relative to the Earth and consequently its gravity is able to stabilize the Earth's rotation. As a result, the Earth has seen less drastic changes in the tilt of its rotational axis than other planets and its climate has been unusually stable. This has facilitated the evolution of life on Earth, especially multicellular and mammalian life, which needs aeons of time to evolve. Moreover, the collision produced the large iron core of the Earth, which generates the strong and stable magnetic field that defends the Earth's atmosphere against erosion by the solar wind. It seems possible that without that collision 4 billion years ago there would be no human beings on Earth.

Mercury
The Late Heavy Bombardment

> Shepherds and people generally are not skilful in sacred
> astronomy, confusing the western and eastern rise. The
> same star may shine in the West when following the Sun
> at a distance great enough to be visible in spite of solar
> splendour, and at another time in the East, when, as
> herald of the day it rises before the Sun, leading it.

Pythagoras, *Timaeus, c.* 360 BCE

Even though Mercury is not far from the Earth, it is something of
an enigma, because its close proximity to the Sun makes it difficult
to observe. We do know that Mercury's surface, like the Moon's,
was heavily bombarded by meteors at an early stage in the planet's
development. But the cause of the bombardment – and what it might
have done to the Earth – is still as much a mystery as Mercury itself.

Mercury is the smallest planet, not much bigger than the Moon.
It is the closest planet to the Sun, which it orbits in eighty-eight
days. Greek astronomers at first had two names for Mercury, as
they thought it was two separate planets, calling it 'Apollo' when
it was to the east of the Sun and 'Hermes' when it was to the west.
It was reputedly Pythagoras who realized that the two bodies were
actually the same planet.

Even though Mercury is relatively near to the Earth, its close
proximity to the Sun makes it very difficult to see using conventional
telescopes. As Mercury is never more than 28° from the Sun, it is
never high in the sky and can only be viewed from the Earth in
the twilight. Nor is observation from space any easier. The Hubble
Space Telescope is not permitted to view Mercury directly, as the
Sun's heat and light could damage it. For much the same reasons,

it is difficult to design a space probe capable of visiting Mercury. A spacecraft will quickly overheat as it approaches so close to the Sun, and will be peppered by storms of solar particles. Furthermore, as the spacecraft drops towards the Sun it will pick up speed as it is drawn down by the Sun's strong gravitational field, and can overshoot Mercury. This must be countered by burning large amounts of fuel or by approaching Venus in just the right direction, at just the right time in the orbits of both planets, so that Venus's gravity can be used to slow the spacecraft down.

Because of these constraints Mercury is one of the least studied planets. Radar has proved a useful tool, but its capabilities are limited at such a distance. In 1965 it was used to show that Mercury turns exactly three times on its axis for every two orbits around the Sun. A solar day on Mercury (sunrise to sunrise) therefore lasts two Mercury years, or 176 Earth days.

The first space probe successfully to visit Mercury was Mariner 10 in 1974–75, but it was more than thirty years before the second space visit was accomplished by the Messenger probe in 2008–9. Mariner 10 discovered that Mercury suffers extreme fluctuations of temperature, ranging from 90 K (-183 °C) to 700 K (427 °C), as the atmosphere is too thin to provide an insulating effect. The floors of some deep craters near the poles never see direct sunlight and never warm above 112 K. There are patches in these craters that seem to reflect radar pulses in the same way that ice does, and specific deposits of ice were confirmed by Messenger. It may have been deposited by melting comets: a large comet impact on Mercury's surface could have generated a temporary steamy atmosphere, which may have condensed and frozen in the dark, cold craters.

The thin 'atmosphere' is constantly being lost and replenished and is therefore properly termed an 'exosphere' rather than an atmosphere. Mercury's exosphere consists mainly of hydrogen and helium picked up from the Sun, but also contains less abundant atoms like sodium and silicon that have been knocked off the surface of the planet by the solar wind, a process called 'sputtering'. A surprise discovery from Messenger was that the exosphere contains water, perhaps derived from the cometary ice at the poles.

Mercury's surface is heavily cratered, like the Moon's. In fact, the oldest surface areas of both bodies were cratered at the same time. To estimate the ages of the surfaces of different planets and satellites, astronomers count the number of meteor craters – the younger the surface, the fewer craters it has, especially large ones. These counts suggest that meteor impacts occurred more frequently throughout the Solar System during an early period in its history. This is confirmed by studies of rocks collected from the Moon by the Apollo astronauts and lunar meteorites collected on Earth.

Lunar meteorites are pieces of the Moon that were knocked off its surface by the impact of asteroids. None of them is older than 3.9 billion years. This posed a puzzle: given that the Moon was 4.6 billion years old, why were the oldest meteorites so much younger? Lunar rocks collected by Apollo astronauts showed that the crust of the Moon had been strongly heated 3.9 billion years ago. What had caused this catastrophic heating? Between 1974 and 1976 studies by a number of planetologists, including Fouad Tera, Dimitri Papanastassiou, Gerald Wasserburg and Grenville Turner, suggested that 3.9 billion years ago asteroids and meteors had heavily bombarded the surface of the Moon for a discrete period of time (200 million years) and melted it. They called this event the 'lunar cataclysm'; it is now known as the Late Heavy Bombardment. The oldest parts of Mercury's surface were cratered during the same period.

No one knows why the bombardment occurred. There may have been a major collision between planets, or a disturbance in the outer Solar System that caused a rain of asteroids or Kuiper Belt Objects to fall inward towards the Sun. The focusing effect of the Sun's strong gravity meant that Mercury was pummelled especially heavily by infalling asteroids. A particularly large impact created the Caloris Basin (plate xv), one of the largest craters in the Solar System, with a diameter of 1,550 kilometres. As discovered by Messenger, the impact caused volcanoes to spring up around the rim of the crater. Mariner 10 had already discovered that shock waves from the impact travelled to the other side of the planet to create an unusual hilly region known as the 'Weird Terrain'. The whole planet would have rung like a bell from the force of the impact, which was on the verge of world-shattering.

Of course, if the Moon and Mercury suffered under the Late Heavy Bombardment, so did the Earth. The Bombardment produced about 1,700 craters on the Moon that were more than 20 kilometres in diameter. Given the larger size of our planet, the bombardment would have generated ten times as many craters on Earth, some of which would have been as large as 1,000 kilometres in diameter. This scenario is supported by the discovery that extra-terrestrial isotopes are especially abundant in deep sediments laid down in Greenland and Canada during the time of the Late Heavy Bombardment. It might also be significant that the fossil record of life on Earth seems to have started after 3.9 billion years ago. If life had begun to evolve before then, the Bombardment would have interrupted the process and erased any earlier traces. Alternatively, the Late Heavy Bombardment may actually have triggered the evolution of life, bringing an abundance of organic molecules to the Earth on asteroids or comets.

The Greenhouse Effect
Venus and the Earth

> The world scientific community has begun to sound the
> alarm about the grave dangers posed...by greenhouse
> warming, and again we're taking some mitigating steps, but
> again those steps are too small and too slow....The runaway
> greenhouse effect on Venus is a valuable reminder that we
> must take the increasing greenhouse effect on Earth seriously.

Carl Sagan, *Cosmos: Who speaks for Earth?*, 1990

The greenhouse effect was discovered on the Earth and on Venus
in parallel. It is a property of some of the gases in the atmosphere
that keeps the surface of the Earth at a comfortable temperature.
The greenhouse effect is essential for life on Earth. However,
man-made (anthropogenic) greenhouse gases threaten to upset
its benign equilibrium. The surface of Venus is hot enough to melt
lead. Could the same thing happen on Earth?

Venus is the second-nearest planet to the Sun, at about three quarters the distance of the Earth. It is Earth-like in its size and physical
properties, such as its internal structure, solid surface, atmosphere
and magnetic field. Seen through a telescope, Venus seems to have an
almost uniform white surface. In 1761 the Russian scientist Mikhail
Lomonosov observed a halo around Venus as it passed in front of
the Sun. In this way, Lomonosov realized that the halo was scattered
light radiating from Venus's upper atmosphere. The planet's white
'surface' was actually an opaque layer of thick white clouds.

Because it is highly reflective and close to both the Sun and
the Earth, Venus is the brightest object in the sky after the Sun
and the Moon. It was thus one of the first celestial objects whose
spectrum (the bands of visible and invisible light it emits) was

photographed. In 1932 Walter Adams and Theodore Dunham, Jr, imaged Venus's spectrum using the 100-inch telescope at the Mount Wilson Observatory in California and specially made emulsions supplied by C. E. Kenneth Mees of the Eastman Kodak Company. They discovered that previously unknown spectral lines in Venus's spectrum were generated by carbon dioxide, which proved that this gas was a major constituent of Venus's atmosphere.

The carbon-dioxide atmosphere of Venus had peculiar properties. Between 1923 and 1928 American astronomers Edison Pettit and Seth Nicholson measured infrared radiation that came from the tops of Venus's clouds and discovered that the temperature there was cold, ranging from -37 to -42 °C. In 1956 pioneer radio astronomer Cornell H. Mayer and his colleagues at the US Naval Research Laboratory measured the microwave radiation emitted from deep within Venus's atmosphere, much nearer to the ground. Mayer's team discovered that the surface of the planet had a very high temperature indeed, over 300 °C, hot enough to melt lead. This was a surprise: the dense and highly reflective atmosphere of Venus should have reduced the amount of solar radiation that was able to reach the planet's surface. In 1961 Cornell astronomer Carl Sagan put forward the currently accepted explanation for the high temperature: the atmosphere of Venus has a strong greenhouse effect, much stronger than the Earth's.

The French mathematician Joseph Fourier first noticed the greenhouse effect in the Earth's atmosphere in 1827. The Earth receives light from the Sun, of which about 70% is absorbed. This sunlight warms the land, atmosphere and oceans, which radiate energy back towards space as infrared light. But most of the infrared emitted from the surface does not escape into space but is absorbed in the atmosphere by greenhouse gases and clouds. It heats the lower air, just as the glass roof of a greenhouse or the windows of a car allow sunlight in but trap the infrared light radiated by everything inside, causing the interior of the greenhouse or the car to become hot.

Throughout the nineteenth and twentieth centuries, scientists gradually came to understand precisely how the greenhouse effect works on the Earth. In 1861 the Irish physicist John Tyndall discovered that water vapour and carbon dioxide are the most important greenhouse gases in the Earth's atmosphere. Astronomer Samuel Langley, director of Allegheny Observatory in Pittsburgh, Pennsylvania, used spectrum mapping in 1884 to show exactly how carbon dioxide and water vapour were opaque to infrared radiation. As Fourier and Tyndall had suspected, these two atmospheric gases allow incoming sunlight to pass through the atmosphere, but block much of the outgoing infrared radiation.

As they learned more about the properties of greenhouse gases, scientists began to wonder whether an increase in the amount of them in the atmosphere might raise the temperature of the Earth. In a series of publications from 1896 to 1908, the Swedish chemist Svante Arrhenius speculated that changes to the level of carbon dioxide in the Earth's atmosphere could alter its surface temperature through the greenhouse effect. Not only did geologically produced changes cause the Ice Ages, Arrhenius thought, but the burning of fossil fuels (such as coal) would produce anthropogenic global warming. Between 1928 and 1938, British meteorologists George Simpson and Guy Stewart Callendar finally succeeded in calculating the full extent of the greenhouse effect on Earth. By 1961 physicists Gilbert N. Plass and Lewis D. Kaplan were making increasingly realistic climate models for the Earth and the phrase 'greenhouse effect' began to be used in connection with global warming. In the mid-1960s the first conferences on global warming were held and the first official reports on global warming appeared.

Scientists were also applying the same ideas to other planets. Between 1908 and 1922, astronomers Frank Very, Charles G. Abbott and Edward A. Milne made various studies of the greenhouse effect in the atmospheres of Venus and Mars, on the basis of existing knowledge regarding their compositions. Astronomer Rupert Wildt even speculated in 1937–40 that the surface temperature of Venus would be exceptionally high 'on account of the greenhouse effect of the carbon dioxide', estimating that large quantities of this gas could raise the surface temperature of the planet by up to 50 °C above what would otherwise be expected. Compared to the Earth, the greenhouse effect on Venus is extreme, so much so that Carl Sagan suggested it had become unbalanced and run away with itself.

Venus's atmosphere was originally carbon dioxide, but the greenhouse effect had warmed the planet so much that the high temperature changed the composition of its surface, generating more greenhouse gases in the atmosphere and raising the temperature still further. Accurate calculations were difficult in Sagan's day because too little was known about Venus in the 1960s. This gap in knowledge was a major impetus for the exploration of Venus by spacecraft.

The US Mariner fly-bys of Venus began in 1962 with Mariner 2. This mission confirmed the high temperature of Venus's surface and the high density of its atmosphere (ninety-three times denser than the Earth's). In 1967 the Soviet Venera spacecraft series, initially designed by pioneering Soviet space engineer Sergei Korolyov, began their scientific explorations of Venus. The Venera missions were the first to venture onto the planet's surface, followed by US Pioneer probes. The spacecraft glided and parachuted through the atmosphere, measuring its properties, but they landed too fast to survive the impact. Later Venera missions between 1967 and 1982 successfully landed softly on the surface, as did two Russian Vega landers in 1985. These missions discovered that Venus's surface was composed of black, scaly volcanic rock under a yellow sky. The surface has a temperature of 740 K (467 °C), compared to the Earth's 287 K (14 °C). Its hot, dense atmosphere is indeed primarily made up of carbon dioxide. Clouds of sulphuric acid droplets are thought to be responsible not only for the yellow sky but also for the intensity of the present greenhouse effect on Venus. Other minor components include nitrogen and water vapour with hydrogen chloride and hydrogen fluoride. Like sulphuric acid, these last two substances are very powerful acids and the landers withstood their corrosive hot rain for only between a few minutes and two hours.

Owing to these severe technical difficulties, no landers have been sent to Venus since 1985. However, it has become possible to maintain orbiters around Venus for longer durations. The Magellan spacecraft lasted four years from 1990 to 1994 and mapped the surface of Venus with radar that penetrated through the clouds. It proved to be a surface of volcanoes and lava flows (plate IV). The space exploration of Venus continued with the European Space Agency's Venus Express, which arrived at Venus in 2005 and identified some regions where volcanic activity was still happening.

Since 1827, when Fourier first identified the greenhouse effect on Earth, carbon dioxide in the Earth's atmosphere has increased by about a third due to man-made emissions, principally from industrial processes; an increase in the amount of methane generated by modern farming practices augments the change. In the absence of man-made emissions, the moderate greenhouse effect on the Earth naturally raises the planet's overall temperature by 33 °C, keeping the climate livable. A doubling of the amount of carbon dioxide by anthropogenic means has the potential to increase the temperature by 3 to 5 °C, causing serious global climate change. However, the runaway greenhouse effect on Venus raises its temperature by 500 °C, completely changing the planet's atmosphere and its climate. Venus is a horrific vision of what catastrophic climate change could look like on Earth.

Mars

The drying, dying planet

> But who shall dwell in these worlds if they be
> inhabited?...Are we or they Lords of the World?

Johannes Kepler, *Harmony of the World*, translated by
Robert Burton, 1618

Why did Mars – the planet in the Solar System most similar to the
Earth – fail to sustain life? Early science-fiction writers portrayed
Mars as a dying planet inhabited by desperate, warlike aliens. We
now know that Mars suffered a global climatic catastrophe early
in its development.

Mars is the red planet, the fourth from the Sun. It is smaller,
colder and drier than the Earth and has a much thinner atmosphere. Galileo saw the disc of Mars with his telescope, but could
not see its surface markings, which were discovered in 1659 by the
Dutch astronomer Christiaan Huygens, who determined that Mars
had a rotation period of about twenty-four hours – its days were
almost the same length as the Earth's. In 1666 the Italian-French
astronomer Gian Cassini discovered Mars's polar caps, which were
assumed to be ice caps like the Earth's; 350 years later, space probes
confirmed that the polar caps are deposits of ice and dry ice, 2 to
3 kilometres thick (plate v).

In 1840 the German banker and amateur astronomer Wilhelm
Beer and his colleague Johann H. von Madler made the first maps
of Mars, showing dark areas that seemed variable in colour and
intensity. Initially these dark areas were thought to be seas, but the
French astronomer Emmanuel Liais suggested in 1860 that they
could be large patches of vegetation, showing seasonal variations
in colour. When Giovanni Schiaparelli mapped Mars in 1877, he

labelled the dark patches as 'continents', 'islands' and 'bays', linked by numerous long, straight *canali* ('channels').

The *canali* led to speculation that Mars was inhabited by intelligent life. The American astronomer Percival Lowell's many maps of Mars (dating between 1894 and 1911) made Schiaparelli's 'canals' straighter and thinner, adding unwarranted credibility to the idea that they were artificial irrigation canals carrying water between darker 'cultivated' areas. Although other astronomers scorned Schiaparelli and Lowell's overly detailed maps as works of fantasy, their vision of Mars took hold in science fiction and popular culture, where the planet was depicted as an old world, inhabited by warlike aliens looking to colonize the Earth because, despite their efforts at irrigation, their own world was dying. But Turkish-born French astronomer Eugenios Antoniadi proposed that the canals were only psychological interpretations of faint, blotchy structures seen through the terrestrial atmosphere. In 1903 the Greenwich astronomer Edward Maunder used schoolboys as test subjects to demonstrate that a defective telescope causes an area with many point-like features (such as a group of craters) to appear as a network of lines.

Mars has two small, potato-shaped satellites. The satellites were discovered in August 1877 by the American astronomer Asaph Hall at the US Naval Observatory during a particularly favourable time when the Earth was unusually close to Mars. Hall's first glimpse of a satellite occurred just before fog from the River Potomac rolled in and shut his observing window. During the ensuing cloudy weather he slept at the observatory so as to take advantage of any brief, clear interval. He found the satellite again a week later, and was so full of his discovery that, in his excitement, he couldn't keep it to himself:

> Until this time, I had said nothing to anyone at the
> Observatory of my search for a satellite of Mars, but on
> leaving the observatory after these observations of the 16th,
> at about three o'clock in the morning, I told my assistant,

George Anderson, to whom I had shown the object, that I thought I had discovered a satellite of Mars. I told him also to keep quiet as I did not wish anything said until the matter was beyond doubt. He said nothing, but the thing was too good to keep and I let it out myself. On 17 August between one and two o'clock, while I was reducing my observations, Professor Newcomb came into my room to eat his lunch and I showed him my measures of the faint object near Mars which proved that it was moving with the planet.

Hall discovered the second moon later that night:

For several days the inner moon was a puzzle. It would appear on different sides of the planet on the same night, and at first I thought there were two or three inner moons, since it seemed very improbable to me at that time that a satellite should revolve around its primary in less time [7 hrs 39 min] than that in which the planet rotates [24 hrs 36 min]. To settle this point, I watched this moon throughout the nights of 20 and 21 August, and saw, in fact, that there was but one inner moon.

The moons were christened after the horses that drew Mars' war-chariot, as related in the *Iliad*: Phobos (the inner moon) and Deimos (the outer) – Fear and Dread. Phobos is spiralling down towards Mars and will impact in 50 million years.

Mars is repeatedly bombarded by meteors. The discovery that Mars is heavily cratered came in 1964 in a fly-by mission by the US space probe Mariner 4. Mars has no tectonic plates, so its surface is not regularly churned over in the same way as the Earth's, and weather erosion of the landscape is minimal because of the thin, dry atmosphere. Much of the crust of Mars is therefore very old; some terrains formed over 3.8 billion years ago and preserve traces of every subsequent meteoric bombardment. There are a few newer terrains on Mars, chiefly ash-strewn volcanoes and lava flows.

Mariner 9 discovered recent volcanic activity on Mars in 1971–72. The largest volcano is Olympus Mons, which at 24 kilometres high is three times higher than Mount Everest. Other dramatic landscapes were shaped by water, including vast, now-dry flood plains and glacial features. Mariner 9 sent back pictures of the eponymous Valles Marineris, a huge canyon system 600 kilometres wide and 7 kilometres deep, which extends 4,000 kilometres east–west along the martian equator.

NASA's Viking missions of 1976 were the first to land on Mars and to image its desert-like surface at close range. The landers looked for and failed to discover organic material in the soil – there did not seem to be life on the desert surface, although it is still possible that evidence may be found elsewhere on the planet. Mars Pathfinder and the Mars Rovers landed in 1997 and 2004, respectively, and Mars Global Surveyor (2001) and Mars Express (2003) mapped the surface at high resolution for several years, confirming that there were massive floods and glaciers on areas of Mars in the past and discovering that water and ice still produce changes on Mars.

The martian atmosphere is made of carbon dioxide, nitrogen and argon. Because it is so thin, it is easy to see clouds, sand storms and seasonal exchanges of material between the polar caps. During the northern winter and southern summer, great dust-storms sometimes cover virtually the whole planet. More frequently, smaller tornados twist across the desert surface as 'dust-devils'. The orbit of Mars changes with time, causing the seasons and climatic cycles to vary dramatically. The martian equivalents of Milankovič cycles are therefore more extreme than those on Earth.

Mars has no magnetic field, but in 1999 Mars Global Surveyor discovered residual magnetism in old surface rocks in the southern hemisphere, laid down as they drifted over the convective iron core of the planet. At some time about 4 billion years ago the liquid core cooled enough to solidify, and the martian dynamo died away. Since Mars does not have a protective magnetosphere like the Earth's, its atmosphere is exposed directly to the solar wind and has gradually been stripped away. This means that ultraviolet light and solar particle radiation reach the surface at levels that would be deadly to surface-dwelling life. The weak atmosphere means that the martian air pressure is only 1% of the Earth's atmospheric pressure, too low for liquid water to remain liquid for long. Without a thick atmosphere to insulate it, the planet does not retain much of the heat it receives from the Sun, causing severe frosts at night with temperatures plunging as low as -140 °C in the polar regions.

These discoveries suggest that Mars was once wet and warm. Its climate changed catastrophically when the planet lost its magnetic field and became the dry and sterile place that it is today. Yet it is still possible that life developed in the wet and warm era and survives in niche environments even now.

Water on Mars and Jupiter's Satellites
Evidence for extraterrestrial life?

A moment of happiness, you and I sitting on the
    verandah, apparently two, but one in soul, you and I.
We feel the flowing water of life here, you and I,
    with the garden's beauty and the birds singing.
The stars will be watching us, and we will show
    them what it is to be a thin crescent moon.
You and I unselfed, will be together, indifferent
    to idle speculation, you and I.
The parrots of heaven will be cracking sugar as we
    laugh together, you and I. In one form upon this earth,
    and in another form in a timeless sweet land.

Mewlana Jalal ad-Din Rumi, 'You and I', fourteenth century CE

Water is the key ingredient for life on Earth. Wherever we find water in the Solar System, it is possible that we may also find evidence of life. 'Splosh' craters, traces of catastrophic floods, glaciers and channels carved by underground rivers suggest that life could at one time – or might even now – exist on Mars. But the most promising places for extraterrestrial life may be on Jupiter's moons, like Europa, which contains more water than the Earth.

When Mariner 9 reached Mars in November 1971 it was the first spacecraft to enter into orbit around another planet. A martian dust-storm was in progress at the time and all that could be seen in Mariner's first transmissions were the south pole and the tops of four high volcanoes. Controllers waited two anxious months before the atmosphere cleared and Mariner began photographing the surface of Mars. By the time it shut down in October 1972 the spacecraft had sent almost seven thousand images back to Earth.

Mariner's success paved the way for the two Viking missions that were launched in 1975. Upon reaching Mars, each spacecraft separated into a lander and an orbiter. The Viking landers showed close-up images of a sandy, wind-blown desert, strewn with angular rocks that had fallen as debris from meteor craters. The Viking missions were followed by Mars Pathfinder in 1997, and in 2004 by the Spirit and Opportunity Rovers, which were mobile and able to range outside their immediate landing areas. Spirit was active until it got stuck in a sand dune in 2010, Opportunity until it was immobilized by a sandstorm in 2018. NASA's Mars Science Laboratory mission used a remarkable crane to lower the Curiosity rover onto the surface of Mars in 2012. Larger and faster than either Spirit or Opportunity, Curiosity has the capability to analyse the composition of rocks at a distance of 7 metres. Another NASA lander, InSight, successfully touched down on Mars in 2018 in order to measure seismic activity on Mars as a means to investigate its interior structure.

Mars Phoenix Lander touched down in 2007 at the most northerly latitude of Mars yet explored. The Viking Orbiters launched in 1975 had mapped the surface in detail from a distance, a task later shared by Mars Global Surveyor (1997–2006), Mars Odyssey (launched 2001), Mars Express (launched 2003), and the Mars Reconnaissance Orbiter (launched 2006). These spacecraft have discovered that the surface of Mars has wide-ranging networks of valleys. Unlike watercourses on Earth, some of these dry riverbeds, which are the remains of an extinct drainage system, have no small streams and tributaries, only large rivers that emerge full-size at their source. There are also numerous glacial features on Mars. It seems likely that these riverbeds were not made by the runoff of rain, but were carved by groundwater flow: rivers flowing at first underground, then emerging from beneath icy glaciers that channelled the water. Other valleys were created when the permafrost above ground was melted by geothermal springs, causing the roof structures of subterranean rivers to collapse.

In addition to its rivers and springs, Mars also had massive floods that lasted for a matter of weeks, producing lakes and even seas. Mars Global Surveyor saw crusty, dried-up lake beds, and stepped cliffs on the interior walls of craters: platforms cut by waves, showing that the craters had once been filled with water. Rovers, like Curiosity, have established the chemical composition of some minerals on Mars that form only in standing water. A startling discovery by Viking Orbiter was of crater-crowned 'islands' standing above a dry plain at Ares Vallis in the Chryse Planitia region. The lozenge shapes of these islands and the fact that their steep cliffs are between 400 and 600 metres high suggests the islands were carved from impact craters by a flood of catastrophic proportions. Mars Pathfinder also saw rounded rocks and boulders of many different compositions that must have been deposited in their present location by floodwater.

Some of these floods were truly awesome in scale. In one case where a natural ice dam had collapsed, something like 100,000 cubic kilometres of water was released in only a few days. In comparison, a typical flood on Earth is only a few cubic kilometres, and the largest known flood in the geological record of our planet released somewhere between 100 and 1,000 cubic kilometres of water.

Melas Chasma is a basin 1,200 metres above the floor of Valles Marineris. It once was filled by a lake and now contains deposits left when the lake dried. Swirling rusty streaks indicate where hardened sediments have been exposed by waves lapping at the edge of the lake.

Where is this martian water now? Some of it is frozen in the polar ice caps (plate XI), but much of it could be hidden under the surface of the planet. At its cold, high-latitude landing site, the Mars Phoenix Lander exposed small quantities of water-ice, which lay millimetres under the surface, by scratching at the ground with a scoop. It is possible that more water-ice is present in a permafrost layer many metres deep. A radar experiment on Mars Express has even suggested that there is a large reservoir of liquid water underground at the martian south pole.

Occasionally, the ice melts. Some recent craters on the surface of Mars, like the one known as Yuty, are surrounded by outward-flowing lobes, which look like the petals of a flower – features not found on the Moon or Mercury. These craters are called 'splosh' craters, and seem to have been formed by projectiles that crashed into mud. This is further evidence that the subsurface soil of Mars may contain ice, which was melted into mud by the meteor's impact, flowed outwards and then resolidified.

The presence of water opens the possibility that life might have developed on Mars at some point in the past. In limited areas of the planet conditions exist that would favour its survival even now. In 2004 astronomers using ground-based telescopes in Hawaii and Chile and the European Space Agency's Mars Express satellite discovered methane in Mars atmosphere, which is released from several vents during the summer, when the ice melts.

On Earth, methane is puffed into the atmosphere by volcanoes and deep ocean vents – but it is also produced by bacteria and animals. On Mars, the Curiosity rover has measured a strong seasonal cycle in the methane concentration. Perhaps this is a consequence of seasonality in the amount of carbon dioxide in the atmosphere: it freezes out onto the large southern polar cap, making the overall atmosphere thinner, which increases the relative concentration of methane, which doesn't freeze. Seasonal variations in dust-storms and the levels of UV light could also affect the abundance of methane. In the back of everyone's mind is the thought that biological activity could well be seasonal. It is too early to say which process is responsible for the methane on Mars.

Mars may not be the only place in the universe where life might once have existed, nor perhaps even the most likely place for astronomers to search for traces of life. There is water elsewhere in the Solar System – lots of it. Jupiter's satellite Europa, which is about the size of Earth's Moon, has been investigated by both the Voyager and Galileo spacecrafts. Europa has a completely spherical, nearly smooth surface with a grooved pattern that looks like crazy paving. Its surface is predominantly water-ice, covered with icy plains. The grooves are cracks in the ice, where floes the size of cities have broken off from the main sheet. The floes are mobile because they float on water. Frozen 'puddles' of ice smooth over older cracks and warmer material bubbles up from below the surface. Evaporative salts tint the white ice a reddish-brown in some areas. The ice layer of Europa could be a kilometre thick, or more. The pressure of the heavy layer of floating ice combines with the radioactive and tidal heating of Europa's interior to liquidize the water beneath the ice. Europa has more water than the total amount found in the oceans on Earth.

Jupiter has two further satellites that may have underground oceans. Ganymede and Callisto have a composition that is a mixture of rock and ice. There is no direct sighting of liquid water but there are several lines of argument that point to both Ganymede and Callisto having oceans of salty water under their rocky surfaces. The ocean within Ganymede is perhaps 1,000 kilometres deep, and like Europa's ocean holds as much as or more water than there is on Earth. Callisto's ocean is only a few hundred kilometres deep.

All these oceans may have life swimming there. Salty oceans warmed under an icy surface: if there is indeed life anywhere else in the Solar System, it could well be here on the moons of Jupiter. Perhaps life in the Universe is rarer on planets than on moons – perhaps it is we who are atypical.

Volcanoes on Io
A chance discovery

> Had the fierce ashes of some fiery peak
> Been hurled so high they ranged about the globe?

Alfred, Lord Tennyson, 'St. Telemachus', 1892 ('suggested by the memory of the eruption of Krakatoa')

'It was a moment that every astronomer, every planetary scientist lives for', recalled Linda Morabito of the instant when she guessed what was causing a mysterious crescent-shaped blob to appear on photographs of one of Jupiter's moons. 'I had the sense that I was seeing something that no one else had seen before.' That evening at dinner, Morabito told her father what had happened. 'He looked at me and said: "Do you realize you may have discovered the first volcanic activity outside the Earth?"'

Jupiter's innermost satellite was discovered in 1610 by Galileo Galilei and named 'Io' by Simon Marius, after one of the Roman god's mythological lovers. Until the advent of large telescopes in the last years of the nineteenth century, Io remained a featureless point, its status as a world or moon 3,600 kilometres in diameter (over a quarter the diameter of Earth) a matter of theoretical inference. In the 1890s, using the Lick Observatory telescopes at the University of California, Edward E. Barnard saw changes in Io's brightness that suggested the colour and reflectivity of its surface varied from area to area (equator and pole). In the twentieth century, R. B. Minton was able to show that Io had reddish-brown poles and a yellow-white equator – its surface colours were very different from the other three ice-covered Galilean satellites. The prevailing theory was that Io's surface was covered with sodium and sulphur salts.

When the Pioneer space probes passed Jupiter in 1973–74, the mass of Io was calculated from the amount by which it deflected the spacecraft off course. It was clear that Io was denser than Jupiter's other satellites, with less ice and more rock. Pioneer 11 took a fuzzy picture as it flew over Io's pole, showing Io's yellow colour and some mysterious dark patches. The next spacecraft to view Io were the Voyager 1 and 2 probes, which flew past Jupiter in 1979. On 5 March 1979 Voyager 1 was only 20,000 kilometres above Io's surface and transmitted magnificent close-up images back to Earth. The surface was brightly coloured with reds, oranges and yellows. There were no meteor craters, showing that the moon's surface was young, and that some form of geological process had erased any craters that had formed earlier. However, Io had pits that resembled calderas, as well as mountains taller than any on Earth and what seemed to be lava flows. It looked like a volcanic landscape. At the press conference immediately after the encounter, the scientists made much of a multicoloured heart-shaped feature that they thought was indeed a volcano. Proof of this would be discovered a few days later: not by an experienced planetary geologist, but by a young navigation engineer processing routine data.

Linda Morabito had begun working at the Jet Propulsion Laboratories when she was a student at the University of Southern California, and by 1977 had been appointed a navigation engineer for Voyager 1. During the encounter with Jupiter, Morabito was working fourteen hours a day in the Voyager navigation area at the control centre where 'data was falling down on us like rainfall and the images were coming in at all hours of the day and night'. Morabito's immediate task during the encounter was to identify background stars in the images from the spacecraft and use them to determine its position so that its trajectory could be corrected in real time. Afterwards the images would be analysed to reconstruct the trajectory as accurately as possible. On the morning of 9 March, after the encounter with Io, Morabito set about this routine analysis. She began processing several images taken by the Voyager 1 spacecraft as it was looking back 'over its shoulder' for one last view of the Jovian system. One image, taken from a distance of 4.5 million kilometres, had been put up on the monitors for everyone to see. Morabito 'stretched' the image – increased its contrast to look for a particular dim star – and noticed something that no one had been able to see on the raw picture: an 'anomaly' to the left of Io, just off the rim of the moon. The anomaly was crescent-shaped and extremely large relative to the overall size of Io.

Careful scientist that she was, Morabito first considered whether the anomaly might have been caused by a piece of debris or a blemish on the camera. When these had been eliminated, only one possible explanation remained – the anomaly had something to do with Io. Io itself was overexposed on the image and it took some work to discover the exact area of Io's surface where the mysterious object had appeared. The area proved to be the large, heart-shaped feature, now called Pele, which had been shown at the press conference. The anomaly was a plume from a volcano, an ash cloud rising more than 260 kilometres above the satellite's surface; the heart-shaped feature was the volcano itself, with its bright orange and green slopes, yellow ejecta and black lava flows, surrounded by a ring of red material ejected over the years and extending up to 600 km from the volcano. Nine such plumes were discovered later in footage from the same fly-by, and a second plume was actually visible in Morabito's discovery image, above a dark patch called Loki, where the volcanic cloud was high above the night side of Io, catching the rays of the rising Sun. The Loki volcano has been continuously active since it was first observed in 1979.

Current estimates suggest that there are at least 400 volcanoes among the hundred or so mountains on Io. The low gravity of Io allows volcanic plumes to reach as high as 500 kilometres above the surface. Although the Galileo spacecraft was prevented from passing too close to Io for fear that the moon's hostile plasma environment would damage it, Galileo nevertheless saw ten plumes in 1998.

Why – uniquely among Jupiter's moons – does Io have volcanoes, and why does it have so many? According to calculations made in 1979 by Stan Peale, Patrick Cassen and R. T. Reynolds the interior rocks of Io are compressed and expanded by the gravity of Jupiter and its other moons. The resulting friction and pressure heat the rocks into liquid magma, triggering volcanic eruptions. The eruptions create pits and mountains on Io's surface, coating it with sulphurous deposits and covering meteor craters with lava flows hundreds of kilometres long. The scale of these flows is enormous: they contain hundreds of times the volumes of lava produced by volcanoes on Earth. Gases released in the eruptions give Io its thin atmosphere, and leak along Jupiter's magnetic field lines, producing striking aurorae. The aurorae are mainly confined to a kinked oval on the cloud tops. Bright spots ('auroral footprints') mark where the magnetic field leads directly from Jupiter's satellites to the top of Jupiter's atmosphere.

Saturn and the Gas-Giant Planets
Lords of the rings

I have been battering away at Saturn, returning to
the charge every now and then. I have effected several
breaches in the solid ring, and now I am splash into the
fluid one, amid a clash of symbols truly astounding. When
I reappear it will be in the dusky ring, which is something
like the state of the air supposing the siege of Sebastopol
conducted from a forest of guns 100 miles one way, and
30,000 miles the other, and the shot never to stop, but go
spinning away round a circle, radius 170,000 miles.

James Clerk Maxwell, letter to Lewis Campbell in 1857

The four outermost planets in the Solar System – Jupiter, Saturn,
Uranus and Neptune – are strikingly different from the terrestrial
planets that are closer to the Sun. They are exceptionally large and
made mostly of gas and ice (plate XIV) rather than rock and dust. When
Galileo first saw Saturn through his telescope, he was astounded to
find that the planet had 'handles'. The 'handles' were, of course, Saturn's rings. Recent discoveries reveal that all the gas-giant planets have
these peculiar systems of rings. But where did the rings come from?

Jupiter, Saturn, Uranus and Neptune have always been much
larger than the rest of the planets in the Solar System. They originally formed from parts of the solar nebula that were too distant
from the Sun for ices to be vaporized by its heat. The gas giants
consequently retained their icy material; this is why they are much
larger than the inner terrestrial planets, whose ices were 'cooked'
away by the newly formed Sun, leaving only small quantities of
rock and dust to form into planets. As a result, the gas giants are
composed principally of hydrogen and helium. They are in this

way like stars, but considerably smaller, so they are not as hot and dense, and do not generate energy through nuclear reactions – you could say they failed to become stars. The interior regions of the two larger gas giants, Jupiter and Saturn, are composed of 'metallic hydrogen', a form of hydrogen created under high pressure that has some of the properties of metallic crystals. Metallic hydrogen was discovered through theoretical analysis and has never been created in the laboratory because the pressures needed are not consistently achievable on Earth (although some scientists are trying hard): Jupiter and Saturn are, it seems, the only places in the Solar System where metallic hydrogen exists.

Each of the gas giants, most visibly Jupiter, acted as a centre for the formation of a miniature planetary system of small planet-like moons. In addition to its satellites, each also has a system of planetary rings, discs of particles that follow nearly circular orbits near its equatorial plane. Saturn's distinctive rings are well known and can be seen with a good pair of binoculars, but the rings of the rest of the gas-giant planets were only discovered in the last decades of the twentieth century.

When Galileo turned his telescope on Saturn, his equipment was not powerful enough to reveal the true nature of the rings. In 1610 he described what he saw as *ansae* (Latin for 'handles'), which he interpreted as large moons. 'I have observed [Saturn] to be triple-bodied. To my very great amazement I saw that Saturn is not a single star, but three together, which almost touch each other.' Two years later the 'moons' had disappeared. 'I do not know what to say about so surprising a case, so unexpected and so novel', he exclaimed. Saturn's rings were edge-on to the Earth at the time, and could not easily be seen. In 1616 he saw a complex shape, because his telescope had improved and the aspect of the rings had again changed. 'The two companions are no longer two small perfectly round globes...but are much larger and no longer round...there are two half ellipses with two little dark triangles in the middle, each contiguous to the middle globe of Saturn, which is always perfectly round.' Galileo was unable to offer a solution to the puzzle.

In 1655 Dutch astronomer Christiaan Huygens took an interest in Saturn and discovered its largest moon, Titan, which lined up with Saturn and its handles. Huygens and his brother produced a more powerful telescope, and in February 1656 Huygens saw that the 'handles' were in fact a ring around the planet. To establish precedence for his discovery, while buying himself time for further study of the ring, Huygens inserted a cryptic remark in his book about Titan, *De Saturni luna observatio nova* ('New Observation of a Moon of Saturn'), that he had found an explanation for the *ansae*, inviting anyone else in the know to step forward. He then described his discovery using an anagram (into which he had put minimal creative effort): aaaaaaa ccccc d eeeee g h iiiiiii llll mm nnnnnnnnn oooo pp q rr s ttttt uuuuu, which he later revealed meant: *Annulo cingitur, tenui, plano, nusquam cohaerente, ad eclipticam inclinato* ('it is surrounded by a thin flat ring, nowhere touching, and inclined to the ecliptic').

In 1787 the French mathematician Pierre-Simon Laplace suggested that Saturn's ring was a set of thin solid ringlets, somewhat like bangles looped around a forearm, because a single solid ring could not orbit around the planet. However, in 1849, the French scientist Édouard Roche calculated that if any solid satellite was in orbit too close to its planet, it would break up under the tidal forces that the planet exerts. The 'Roche limit', within which a solid satellite cannot survive, is 2.44 times the radius of the planet, a little farther than the external radius of Saturn's rings. This made it unlikely that Saturn's ring could be a solid structure. In 1857 the Scottish physicist James Clerk Maxwell showed that 'the only system of rings which can exist is one composed of an indefinite number of unconnected particles, revolving round the planet with different velocities according to their respective distances.' This was confirmed in 1895 when Allegheny Observatory director James Keeler found that the inner part of the ring was orbiting Saturn faster than the outer part.

As telescopes improved in the seventeenth and eighteenth centuries, astronomers saw that Saturn's ring was in fact a series of rings.

In 1675 Cassini saw two main rings, which were later named 'A' and 'B', separated by a gap now called the Cassini Division. In 1850 the Harvard Observatory father-and-son team William and George Bond discovered a third ring inside A and B, called the 'crepe' or C ring. Keeler discovered in 1888 a 325-kilometre gap in the A ring, which he named the Encke Gap after the German astronomer Johann Franz Encke, who had studied the rings. The gap is actually the orbital path of a small moon, Pan, which sweeps aside or accretes the particles in the rings as it orbits Saturn. The particles are snow-like ice fragments. The accreted particles fall slowly on the equator of Pan and have built up an equatorial wall of snow. Pan has the shape of a flying saucer, or more domestically, a piece of ravioli.

The Pioneer 11, Voyager and Cassini spacecraft discovered further, thinner rings around Saturn, and a marvellously detailed structure within the rings themselves. The total mass of Saturn's rings is approximately equivalent to the mass of its satellite Mimas, which fits with the theory that the rings are pulverized remains of a former moon. They consist of fragments 1 centimetre to 5 metres in size (plate xx), made primarily of water-ice, or of micron-sized particles. The beautiful structure of the rings is the result of the complex pulls of Saturn's larger satellites which, like Pan, act as 'shepherds', ushering particles out of some zones to leave gaps, much as a shepherd's dog might chivvy sheep to leave one area of a field for another.

The planet Uranus has more than ten rings. The first of these rings was discovered when Uranus passed across a star by chance in 1977. The telescopes of the Gerard P. Kuiper Airborne Observatory, a modified C-141A towing aircraft flying at stratospheric altitudes, were preparing to observe the way that the star's light faded as it passed behind Uranus's atmosphere. But before the planet reached its predicted position in front of the star, the starlight unexpectedly dimmed several times. It had been temporarily blocked by Uranus's planetary rings. After the star exited from behind the planet, it was blocked again by the other side of the rings, with the depth of the dimmings repeating in reverse order. The Voyager 2 spacecraft directly imaged Uranus's rings in 1986.

After the discovery of Uranus's rings, it seemed likely that Neptune might also have rings. Using the same technique, astronomers carefully observed stars for signs of premature dimming as they were occulted by Neptune, and in the 1980s found the evidence they were looking for. To confirm this discovery, NASA engineers reprogrammed the orbit of Voyager 2 as it approached Neptune in 1989, both to learn more about the ring system and also to avoid the risk that dust from the rings would endanger the spacecraft. The Voyager images showed that the planet has four, perhaps five, dusty narrow rings, which have since been named Adams, Le Verrier Lassell, Arago and Galle for the astronomers involved in the discovery of Neptune. The Adams ring is incomplete and has three main arcs named Liberté, Égalité and Fraternité.

Jupiter's rings were discovered by the Voyager spacecraft in 1979, and further studied by the Galileo mission. They are very thin and faint but have been imaged by the Hubble Space Telescope.

Rings of particles and debris do not orbit a planet forever, but fade away – relatively quickly, compared to the age of the Solar System. Saturn's rings may have been caused by the break-up of a 200-kilometre satellite less than 500 million years ago; other planets' rings formed from fragments of small icy satellites or captured comets, and some are periodically replenished by sprays of dust from meteoroid impacts on rocky satellites. Our own planet almost certainly had a ring system at least once in its history, after the Moon was formed. The rings would have presented a beautiful spectacle in the sky, whether they were seen only by the uncomprehending eyes of dinosaurs, or whether no creatures had yet evolved that were able to see them at all.

The close scrutiny given by astronomers to Saturn's rings revealed its satellites. Christiaan Huygens discovered the first, Titan, in 1655. It is the second-largest moon in the Solar System, just 100 kilometres smaller in diameter than Jupiter's Ganymede.

From Earth Titan is inscrutable, covered by an impenetrable hazy atmosphere. It is the only moon in the Solar System with a thick atmosphere. It is primarily composed of nitrogen. The haze arises because the atmosphere of Titan also contains methane. Sunlight acting on methane produces particles that are akin to smoke: the haze is 'smog'.

Titan has been examined from close quarters by an ESA probe named Huygens in tribute to Titan's discoverer. Carried to the moon by NASA's Cassini spacecraft, Huygens parachuted through the atmosphere and landed on Titan's surface in 2005. As it descended, it pictured its landing area, a flat plain adjacent to hills cut by river valleys. It touched down in the damp estuary of one of the valleys, squelching onto a lake bed littered with boulders. It looked like a mundane landscape, but was literally and metaphorically out of this world. Although Titan's landscape shows similarities to any landscape sculpted by water, the liquid concerned is not water but liquid methane. Methane rain falls on the hills, running into methane rivers that flow down to methane lakes. The lake where Huygens landed was almost dry because of seasonal effects on Titan, and is one lake among many: the lake scenery over Titan was mapped soon afterwards by Cassini as it orbited in the Saturnian system. A radar system on Cassini probed below the haze to view the flat lake surfaces.

The atmosphere of Titan is like the Earth's atmosphere in its early years, pre-biotic, before life evolved here and photosynthesis produced oxygen, which chemically combined with methane and similar molecules to produce the atmosphere in which we live today. Titan offers us a glimpse of how the Earth used to be.

THE DYNAMIC

UNIVERSE

Helium

The cosmic element

> On the subject of stars, all investigations which are not
> ultimately reducible to simple visual observations are
> necessarily denied to us. While we can conceive of the
> possibility of determining their shapes, their sizes, and their
> motions, we shall never be able by any means to study their
> chemical composition or their mineralogical structure.

Auguste Comte, *Cours de la Philosophie Positive*, 1835

The discovery of helium in 1868 was a transformative moment for
chemists and astronomers, entirely disproving Comte's notion that
the stars were inherently unknowable. The revelation came as a faint
yellow line of light, observed during an eclipse of the Sun. It was
emitted by a major ingredient in the makeup of stars, which was
later found also to be an important building block of substances
on Earth. Copernicus and Galileo were right: the Earth and the
heavens were made of the same basic materials.

In the Middle Ages, astrology – the arrangement of the planets
in the zodiac – was part of the study of alchemy, an early form of
chemistry whose original purpose was to turn base elements into
precious metals. The relationships that alchemists perceived between
planets and chemicals were reflected in the old names for certain
metals. Quicksilver is still called mercury, but other cosmic names
have fallen into disuse: copper was once known as Venus, iron as
Mars, tin as Jupiter, lead as Saturn, gold as the Sun, and silver
as the Moon. As recently as a century ago, household pipes were
still stamped with the astrological symbol for the planet Saturn to
indicate that they were made of lead.

Eventually alchemy evolved into the modern science of chemistry.
The French chemist Antoine Lavoisier, in a series of experiments

conducted between 1782 and 1789, discovered by carefully weighing his chemical compounds before and after they participated in reactions that some chemicals were never broken down into lighter ones. He called these chemicals 'elements', publishing in 1789 a list of thirty-three elements (although not a list that modern chemists would entirely agree with). Through the work of nineteenth-century Russian chemist Dmitri Mendeleev, this list evolved into a precursor of the modern Periodic Table, with the known elements grouped into columns and rows according to their observed chemical properties (which we now know are dictated by their atomic structures).

The empirical arrangement of the elements in the early Periodic Table left a number of empty holes, sparking a search for the missing elements, many of which were subsequently named for celestial objects, reflecting the traditional association of chemicals with planets. Uranium was named for the planet Uranus in the eighteenth century; a few years later, palladium and cerium were named after the recently discovered asteroids Pallas and Ceres. Neptunium and plutonium were named for the planets Neptune and Pluto, tellurium from the Greek word for the Earth and selenium for the Moon. Some of the celestial names that were given to the new elements are no longer used, including aldebarium and cassiopeium, names derived from the star Aldebaran and the constellation Cassiopeia, for the elements now called ytterbium and lutetium, respectively. Denebium (from the star Deneb) was a name given to a rare earth element whose existence was later disproved.

Such associations between newly discovered elements and the cosmos were only products of a poetic and fanciful naming convention. The first clues that the terrestrial elements were actually to be found elsewhere in the cosmos came from the spectrum of sunlight.

In 1802 William Wollaston discovered that the spectrum of sunlight had seven gaps, which he regarded as boundaries between the natural colours of the spectrum. But in 1814 the German optician Joseph von Fraunhofer invented a spectroscope with superior resolution and discovered not seven but hundreds of gaps in the solar spectrum. The gaps are now known as the 'Fraunhofer lines'. Fraunhofer accurately measured their wavelengths (as an aid to making accurate optical instruments) and labelled the more prominent gaps with letters. The German chemists Robert Bunsen and Gustav Kirchhoff discovered that many of the Fraunhofer lines represented light that had the same wavelength as the light that materials emitted when they were heated and vaporized. For example, the Fraunhofer D-lines in the solar spectrum were identical with the yellow sodium emission from salt. This suggested that sodium was a component of the material heated in the atmosphere of the Sun.

By the end of the 1880s, spectral emissions from fifty of the then known elements had been discovered in the solar spectrum. This proved that the Sun was made of similar elements to the Earth. Applying spectroscopy to the brighter stars, Henry Draper and William Huggins showed that stars also had dark lines in their spectra. The spectroscopists Father Angelo Secchi, H. C. Vogel and E. C. Pickering developed schemes for classifying the spectra of stars and listed the Fraunhofer and other spectral lines that were found in each. The same elements that had been found on the Earth were present not only in the Sun, but also in the stars.

In 1868–69 there was another dramatic development. Astronomers Norman Lockyer and Jules Janssen observed the solar chromosphere (the Sun's denser, lower atmosphere) during the total solar eclipse of 1868 and discovered a strong spectral emission at a wavelength near to the sodium D-lines. To make it practical to measure the wavelength of this light, Lockyer and Janssen developed a technique for viewing the spectrum of the chromosphere in the absence of a solar eclipse. Measured at leisure in the solar observatory, the wavelength of the light emitted by the Sun's lower atmosphere was proved to be different from the sodium D-lines. In fact, the light did not correspond to the emissions from any of the then known elements.

Janssen and Lockyer realized that they had discovered a previously unknown element. Lockyer named it 'helium', after the Greek word for the Sun, *helios*. This element was isolated on Earth in 1895 by the Scottish chemist William Ramsay as he studied radioactive minerals that give off helium as they decay. Helium is thus unique among the elements in the periodic table, having been discovered in a cosmic object before it was identified on Earth. Most terrestrial helium is made by the radioactive decay of heavier elements on Earth – the same process Ramsay observed in his experiments – but most cosmic helium originated during the Big Bang, or is generated inside stars.

In the wake of the discovery of helium came many similar claims for new elements, most of which turned out to be cases of mistaken identity. As conditions on the Sun are so different from conditions in a laboratory, its spectrum is easy to misread. New spectral lines were discovered in the chromospheric spectrum during the total solar eclipse of 1869, but the new element, 'coronium', invented to account for them, was shown in 1941 by the astronomer Walter Grottrian to be nothing more than iron heated to high temperatures and low densities. Similarly, some spectral lines in nebulae were once attributed to the element 'nebulium', but turned out to have been produced by oxygen and other common elements.

In both the discovery of helium and the rectification of spurious claims for other new elements, scientists used new astronomical discoveries to explain terrestrial phenomena, and vice versa. This was an extension of Copernicus's realization that the Earth was a planet like others: the material that the Earth is made of is the same as the material of the rest of the Universe, and the scientific laws that apply here are the same as those that apply everywhere.

Gravitation
Determinism and chaos

> Truth is ever to be found in the simplicity, and not
> in the multiplicity and confusion of things.
>
> Isaac Newton, *Fragments from a Treatise on Revelation*, 1680s

When the theory of gravitation emerged in the seventeenth century it seemed that mathematics had infinite power to see the future. But Newton could only predict the behaviour of one small planet orbiting a lone, absolutely spherical star. The real Universe has many planets, stars and irregular shapes, and by the twentieth century chaos reigned supreme.

Between 1609 and 1621, Johannes Kepler formulated three laws of planetary motion, which he derived from the accurate measurements of the orbit of Mars made by his teacher, Tycho Brahe. In his First Law, Kepler found that the path of each planet around the Sun is not perfectly circular, but elliptical, with the Sun at one of the foci of the ellipse. He formulated a further two mathematical laws, describing the rate at which each planet moves along its orbit as dependent on the size of the orbit. The Second Law was that the line joining the Sun to the planet sweeps through equal areas of the orbit in equal periods of time. Finally, the Third Law was that the ratio of the squares of the orbital periods for two planets is equal to the ratio of the cubes of the semi-major axes (half the longest diameter of each orbital ellipse).

In his 1687 treatise *Principia*, Isaac Newton showed that Kepler's laws were underpinned by the more fundamental theories of dynamics and gravitation, and added an all-important principle about gravity to existing theories of dynamics: all bodies in the Universe attract one another across space, and the force of this attraction between any two bodies varies according to the inverse square of the distance between them (that is, 1 divided by the distance squared). The popular story of this discovery is that in 1666 Newton was musing in the garden of his mother's house in Lincolnshire when he saw an apple fall. He began to think that the power of gravity was not limited to a certain height above the ground but extended to the Moon and beyond.

The first record of the anecdote dates from the year before Newton died and it may be the poetic reminiscence of an old man, a story improved in the retelling; the echoes of the Genesis account of the Tree of Knowledge add to its repeatability. Whatever the circumstances, Newton had a flash of insight about gravity that sparked his investigation into how the Moon orbited the Earth and the Earth the Sun. His findings eventually would transform every branch of science and alter the fundamental human view of the Universe.

Newton's assertion that objects attracted each other across space was controversial. The concept was hard for everyone to grasp and was ridiculed by philosophers, particularly Cartesians, who thought that space was filled with a substance called the plenum, swirling in vortices that transmitted force from one body to another. But Newton's theory of gravitation worked. It explained the motions of the planets and even the shape of the Earth. It enabled Edmond Halley to predict the return of his eponymous comet, and in due course led Urbain Le Verrier and John Couch Adams to the discovery of Neptune in 1846.

Modern astronomers use Newton's theory not only to calculate the motions of planets in the Solar System but also to calculate how satellites move in orbit. These calculations are essential in planning complicated orbital tours that loop a spacecraft via nearby planets to more distant ones using the 'gravity-assist' technique to make the spacecraft pick up (or lose) speed – for instance, when approaching the planet Mercury, which is perilously close to the Sun and notoriously difficult to reach. Although computers enable more complicated calculations to be made accurately, the essential theory has remained the same for over 300 years.

Newton's theory of gravitation was considered superior to Kepler's laws of planetary motion, which only apply to cases where two bodies are interacting (usually the Sun and a planet, but also two stars or two galaxies). A comet may orbit through the Solar System controlled mainly by the Sun and obey Kepler's laws, but it may pass close to a planet and be pulled from its elliptical orbit, at which point Kepler's laws fail to predict its movements because three bodies are involved in the interaction. Moreover, the orbits of planets are actually more complex than the simple ellipses assumed by Kepler – for instance, there are perturbations in the Earth's orbit that cause the Milankovič climate cycles. In principle, Newton's theory could be applied to any type of orbit and any number of planets, stars and galaxies.

In practice, however, the extension of Newton's theory from two bodies to even just three proved difficult, indeed, intractable. Entering an 1887 competition to solve what by then had become known as the 'Three-Body Problem', the French mathematician Henri Poincaré found that he could not give an exact prediction for the orbits of three stars or planets mutually attracted by gravity. He was able to calculate the orbits numerically – we would nowadays do this by computer, he did it by hand – but the paths were 'so tangled that I cannot even begin to draw them'. Moreover, he found that when the three bodies were started from lightly different initial positions, the orbits would be entirely different. 'It may happen that small differences in the initial positions may lead to enormous differences in the final phenomena. Prediction becomes impossible.' Poincaré had discovered a concept that we now term 'chaos theory'.

Poincaré's work has been confirmed by modern computer techniques. The planetary orbits, especially those of the inner planets, are 'chaotic'. If you displace one of the planets by just a single centimetre from its initial position, you might logically expect to find the same single centimetre difference in the planet's final position in 10 million, or even 100 million years. But in practice the planet's final position could be anywhere in its orbital range.

In modern physics, 'chaos' is the word used to describe behaviour that is predictable in the short term but that in the long term depends so much on the starting conditions that minuscule changes can have enormous effects that are impossible to calculate. Weather can be predicted, more or less accurately, one day or one week ahead. However, since something as negligible as a butterfly flapping its wings can set off long-term changes in the air currents, and since there are millions of butterflies all over the world constantly flapping their wings, meteorologists cannot predict at any given moment whether a hurricane will strike Texas a year later. This phenomenon was discovered in 1963 by Edward Lorenz, an MIT meteorologist. He was testing a new computer and re-ran a weather model that he had run before on the old computer. For simplicity in running the test, he included fewer decimal places in the initial data. When he ran the simulation again, he found that the weather patterns generated were completely different. The marginal changes in the data simulated the uncertainties of the real measurements, so it was not a problem with his model, nor with his new computer: it was an innate practical limitation of mathematics. It showed why weather prediction was such an uncertain business. Lorenz called the problem the 'butterfly effect'; later, American mathematician and physicist James Yorke coined the name 'chaos'.

In principle, Newton's theory of mutual attraction could be used to calculate the future state of the entire Universe. The French mathematician Pierre-Simon Laplace therefore imagined a demon who would be able to predict everything that would happen, down to the movement of the tiniest atom:

We may regard the present state of the Universe as the effect of its past and the cause of its future. An intellect which at a certain moment would know all forces that set

nature in motion, and all positions of all items of which
nature is composed, if this intellect were also vast enough
to submit these data to analysis, it would embrace in
a single formula the movements of the greatest bodies
of the Universe and those of the tiniest atom; for such
an intellect nothing would be uncertain and the future
just like the past would be present before its eyes.

This was the first published articulation of scientific determinism. But because the Universe is so large and made up of so many
bodies and particles, Newton's theory – for all the astounding
discoveries it has made possible – cannot predict the future. The
ultimate secret of the Universe is still a secret.

Relativity

The nature of space and time

> Spacetime tells matter how to move; matter
> tells spacetime how to curve.
>
> J. A. Wheeler, *A Journey into Gravity and Spacetime*, 1990

According to Galileo and Isaac Newton, who followed the classical Greek philosophers, space and time are separate from each other, and together form a framework within which events occur. But Albert Einstein thought space and time were more closely connected. He saw them as a single entity: spacetime. This spacetime is not simply the framework or stage on which events unfold – it affects how they unfold. Einstein's theory would inform all major twentieth-century discoveries about the nature of the Universe.

Einstein's theory of Special Relativity is based on two principles. The first principle is that every law of nature has the same mathematical form. The second principle is that the speed of light (represented by the letter c in Einstein's equations) is the same to all observers who move at constant velocity relative to one another.

Relativity is a more important concept in astronomy than in everyday life. For example, because the distances between places in space are enormous, light takes years, perhaps millions of years, to move between them. Two events that are simultaneous for one observer may therefore occur at different times according to other observers. A double-star eclipse in the Milky Way and the explosion of a nova in another galaxy, which on Earth we see happening in the sky on the same night, may appear at completely different times for observers in another, distant galaxy.

Time dilation is another effect of relativity that has immediate astronomical consequences. The faster an object moves, the slower it seems to experience time – at least to the observer. A clock that is moving (for instance, aboard a spacecraft in orbit) runs slow when compared with an identical clock that is stationary on the ground. Terrestrial clocks are used to time the ticks of pulsars; however, terrestrial clocks are not actually stationary, but run fast and slow depending on the speed of the Earth as it revolves around the Sun in its eccentric annual orbit. This time dilation – a difference in the measurable passage of time due to an object's physical motion – must be taken into account when terrestrial clocks are used to measure astronomical processes. The effects of time dilation are not limited to man-made terrestrial clocks, but apply to any physical process of fixed duration. Supernovae have a 'clock', namely the time that it takes for their light to fade from maximum brightness. Because of time dilation, distant supernovae travelling at high speeds inside their parent galaxies, which are receding further from the Earth as the Universe expands, fade more slowly than nearby supernovae in our own Galaxy, which are travelling at the same speed as the Earth.

In the case of cause-and-effect events, however, the signal that triggers the effect cannot travel faster than the speed of light. As a result, the cause always precedes the effect, no matter how fast you are travelling when you see the two events. This rule is used in astronomy to estimate the maximum size of a source of variable light, such as a pulsar. If a star varies in intensity during a clearly defined period of time (T) – say, 1 minute – its size cannot be greater than cT (that is, 1 minute multiplied by c, the speed of light, or 1 light minute). This is because different parts of the star must be able to communicate with each other so that their brightnesses vary together. This argument was used to discover the small size of quasars, because some vary over only a day and therefore must be less than a light day (about the size of the Solar System) at their maximum dimension; pulsars vary in less than a second and therefore must be less than a light second in diameter – they cannot be bigger than a planet.

The relationship between mass and energy in the theory of Special Relativity is fundamental to understanding the nature of stars. When a body's mass reduces by a given amount (represented in equations by m), its energy (represented by E) reduces by an equivalent amount. This is the iconic $E = mc^2$, equation, which was first presented by Einstein in 1905.

The equation expresses the direct relationship between an object's mass and its energy that underlies the whole of nuclear physics. The speed of light, c, is a big number, and c^2 is even bigger, so a little mass makes a lot of energy. The equation explains why huge amounts of energy are released when hydrogen fuses to helium, and powers the stars, and why gravitational waves have been detected from the huge amounts of energy liberated by merging black holes, even though the gravitational waves are so weak.

Einstein's theory of General Relativity represents a finetuning of the work of Galileo and Isaac Newton. It incorporates the Principle of Equivalence discovered by Galileo, which Newton had incorporated into his theory of gravitation.

Galileo knew that the gravitational force on a body is directly proportional to the mass, and that the body's resistance to being moved by gravity (physicists call this 'inertia') is also proportional to the mass. The Principle of Equivalence claims that these two factors precisely cancel each other out. This is why all bodies falling in gravity fall together, no matter what their mass. Galileo is said to have dropped two weights of different sizes from the Leaning Tower of Pisa to demonstrate this: in the Earth's gravity both weights fell at the same rate and hit the ground at the same time, so far as he could judge. The Apollo astronaut David Scott repeated the experiment on the airless Moon in 1971, when a feather dropped to the lunar surface without air resistance hit the ground at exactly the same time as a hammer.

Newton's theory of gravity works well enough for most calculations of the orbits of the planets but on rare occasions – for instance, in the case of the planet Mercury – its accuracy subtly and mysteriously fails. Einstein discovered the reason for this in 1915. He realized that General Relativity alters the orbit of Mercury by an extra 43 arc seconds ($^{43}/_{3600}$ of a degree) per century because of its close proximity to the Sun, which exerts a strong gravitational force. This discrepancy had confounded astronomers since the nineteenth century. The French astronomer Urbain Le Verrier had thought it might be caused by an undiscovered planet, which he called Vulcan and unsuccessfully tried to locate inside Mercury's orbit. Vulcan faded from astronomy. When Albert Einstein discovered that the discrepancy could be explained by his theory of General Relativity, he reported that 'for a few days I was beside myself with joyous excitement'. The realization that General Relativity could explain the longstanding mystery of Mercury's orbit gave him confidence to publish his theory.

General Relativity has, since then, facilitated countless astronomical discoveries. General Relativity is used to explain how black holes work, and to calculate the orbits of the binary pulsars, which test the theory to the limit. In the twenty-first century General Relativity is essential in solving the problems of dark energy and gravitational waves.

Radio Waves
A new window on the Universe

> How fathomless the mystery of the Unseen is! We cannot
> plumb its depths with our feeble senses – with eyes which
> cannot see the infinitely small or the infinitely great, nor
> anything too close or too distant, such as the beings who live
> on a star or the creatures which live in a drop of water...Ah!
> If we had other senses which would work other miracles for
> us, how many more things would we not discover around us!
>
> Guy de Maupassant, *Le Horla et autres contes fantastiques*, 1887

Radio waves are a type of invisible radiation at the extreme end
of the electromagnetic spectrum. Stars and galaxies emit radio
waves of various lengths, along with the full spectrum of visible and
invisible light. When a radio engineer in New Jersey heard a faint,
mysterious static on his home-built antenna, a window began to
open on this unseen universe.

In 1928 radio engineer Karl Jansky started work at Bell Telephone
Laboratories in northeast New Jersey. His job was to investigate
the sources of interference that might affect transatlantic telephony,
such as electrical equipment and automobile ignition systems, but
also natural sources, such as thunderstorms. To this end he built
an antenna that was sensitive to emissions at a wavelength of 15
metres (the sort of radio waves that you can pick up on a shortwave
radio) and had a certain amount of directional discrimination.
The antenna consisted of an open rectangular wooden frame with
aerial wires strung over it, which rotated around a track on wheels,
earning it the name 'the Merry-Go-Round'.

By 1932 Jansky had found three natural sources of 'static', or
radio noise. The first type of interference was clearly associated with

local thunderstorms. A second type had similar characteristics but was weaker and steadier, with occasional peaks. He realized that this static corresponded to distant thunderstorms in the tropics, and that it reached New Jersey via radio waves that bounced off the ionosphere (the layer of electrically charged plasma in the Earth's upper atmosphere). But the third type of natural static was a mystery. Jansky described it as 'a very steady hiss-type static'. It was so weak that it had little practical effect on radio telephony, but Jansky's curiosity was aroused and he decided to investigate it further.

As he rotated the Merry-Go-Round, Jansky noticed that the mysterious static at first seemed to peak in intensity when the antenna was directed towards the Sun, but as the year progressed this correspondence broke down. Jansky began to study astronomy textbooks and concluded that the source of the hiss was not the Sun, but an unknown object fixed in space. In 1933 he presented a paper to fellow radio engineers called 'Electrical Disturbances Apparently of Extraterrestrial Origin'. A press release on the subject by Bell Laboratories led to immense publicity, but astronomers who picked up the discovery puzzled fruitlessly over the precise origin of the static, and it was Jansky himself who discovered that the steady hiss came from a band along the Milky Way, peaking in the constellation Sagittarius, which is the location of the centre of the Galaxy. However, the economic stress of the Great Depression put a stop to Jansky's research and forced him to turn to work of more practical benefit to his employer.

Grote Reber, a radio engineer who pursued astronomy as a hobby, was one of the few of Jansky's colleagues who continued to investigate the static. In a suburban lot in Wheaton, Illinois, Reber built a parabolic reflector 9 metres in diameter, which was able to measure the strength of the mysterious radio emissions from the Milky Way at metre and centimetre wavelengths. Reber's reflector attracted much curiosity; when a light plane circling the radio telescope suffered an engine failure and had to make an emergency landing, a rumour circulated that he was transmitting a death ray. Reber was the first to map the Milky Way using radio waves. Like Jansky, he found a peak in Sagittarius at the position of the Galactic Centre, but also saw other bright radio sources in Cygnus and Cassiopeia. They were not accurately enough located to identify by Reber's pioneering work but proved to be an exploding galaxy and a supernova remnant. Later, Reber pioneered investigations into very long-wavelength radio astronomy, and emigrated to Tasmania, where, because of the island's position in the magnetic field of the Earth, it is easier to study this type of radiation.

Radio astronomy grew rapidly following the Second World War, aided by the significant wartime investment in radar technology. In Britain, radio-astronomy groups grew up at the universities of Cambridge and Manchester, manned by engineers who had worked on radar. The Lovell Telescope was built by Bernard Lovell near Manchester in 1957, at the time the largest single-dish radio telescope. Britain was a pioneering nation in radio astronomy, its cloudy skies offering no disadvantage. The same refocusing of attention by engineers from wartime radar to astronomy happened in Australia and the USA. The Sun was shown to be a strong radio source – this had first been detected during the War, but was kept secret at the time in case the enemy exploited the knowledge by launching attacks when the Sun was active and confusing radar.

The Milky Way's radio emission proved to be of two kinds: some is diffuse emission from interstellar space – electrons releasing radio waves as they gyrate around the Galaxy's magnetic field; other radio emissions come from isolated sources in the Galaxy. The strongest radio source in the constellation of Taurus, called Taurus A, was the first to be identified: it was actually the Crab Nebula, a supernova remnant. Another radio source, Cygnus A, proved to be a galaxy far beyond the Milky Way – we now know that it has an active nucleus (a black hole). Such galaxies are called 'radio galaxies' because of their strong emissions; analysed collectively as surveyors' markers that populated expanding space, they provided the first proof that the Universe was truly evolving and had an origin in a Big Bang.

Jansky's discovery made it possible to view the Universe outside the spectrum of visible light. It opened up the study of astronomical objects that were previously invisible and drew attention to spectacular objects that, in visible light, had looked unremarkable. Astronomers realized that it could be equally profitable to use other radiations – such as infrared, X-rays or ultraviolet light – to explore the Universe, and began developing new kinds of telescopes, detectors and space vehicles. It was like opening a window inside a house and seeing, for the first time, expansive views of an entire new world beyond the four walls.

X-Rays from Space

The energetic Universe

> The theoretical predictions did not provide much
> encouragement. While several 'unusual' celestial objects
> were pinpointed as possible or even likely sources of X-rays,
> it did not look as if any of them would be strong enough to be
> observable with instrumentation not too far from state of the
> art. Fortunately, we did not allow ourselves to be dissuaded.
> As far as I am personally concerned, I must admit my
> motivation for pressing forward was that I have a deep-seated
> faith in the boundless resourcefulness of nature, which so
> often leaves the most daring imagination of man far behind.

Bruno Rossi, *X-Ray Astronomy*, 1974

X-rays make it possible to see stars and galaxies as well as broken bones and the contents of a suitcase at an airport security check. Like radio waves, X-rays are a type of invisible light given off by stars and other space objects, but they can only be studied at exceptionally high altitudes, using detectors mounted on rockets or satellites. X-ray astronomy is therefore one of the great achievements of the space age – another new window opening onto the Universe.

X-rays lie between ultraviolet light and gamma rays on the electromagnetic spectrum. The Earth's atmosphere completely absorbs X-rays before they reach the ground, so X-ray astronomy is exclusively a space activity. X-rays from the Sun were first measured by the American scientist Herb Friedman in the 1940s, using a Geiger counter mounted on a V-2 rocket that had been captured from Germany at the end of the Second World War. The experiment was organized by the Naval Research Laboratory (NRL) in Washington, DC, as part of a programme to discover how the ionosphere affected the propagation of radio waves. There is a very hot outer layer in the Sun called the corona, and the director of the research programme at the NRL, Edward O. Hulbert, suggested that X-rays from the Sun's corona produced the Earth's ionosphere.

The first attempted observation of solar X-rays on 28 June 1946 may actually have been successful, but the data could not be retrieved due to a hitch, which had surprisingly been unforeseen: the rocket carrying the equipment re-entered the atmosphere at supersonic speeds and buried itself 10 metres into the ground, smashing the detectors to pieces. In later experiments, the instruments were moved to the tail of the rocket and jettisoned before impact. In September 1949, Friedman was finally able to prove that the Sun was indeed a source of X-rays.

The Sun emits X-rays from its hottest active regions. The amount of activity on the Sun varies over an eleven-year cycle, as observed in a dramatic sequence of images obtained by the Japanese Yohkoh X-ray astronomy satellite between its launch in 1991 and its destructive re-entry into the Earth's atmosphere in 2005.

The launch of Sputnik 1 by the then USSR in 1957 provoked the USA to expand its space programme, and several organizations, including NASA and American Science and Engineering (AS&E), were founded to carry out space research. One early AS&E recruit was an Italian particle physicist, Riccardo Giacconi. Following a suggestion made in 1959 by Bruno Rossi, an influential scientist in the US space programme, Giacconi turned his attention to X-ray astronomy – a field that was then completely empty of objects to study except the Sun. Mindful of the exciting astronomical discoveries that had recently been revealed by radio waves, Giacconi and Rossi suspected that X-rays had similar potential.

Teaming up with fellow scientists at AS&E, including Herb Gursky, Bruno Rossi and Frank Paolini, Giacconi developed X-ray detectors and telescopes, and persuaded the US Air Force that it was worth investigating whether X-rays came from the Moon. The team already knew that the Moon is cold and emits no X-rays of its own accord, but they suspected that streams of solar particles might hit its surface and produce X-rays, a natural manifestation of what had been a repeated laboratory experiment. The intention was to use the Moon to monitor the flow of particles from the Sun. The US Air Force made large Aerobee rockets available at its White Sands launch site in New Mexico, and in June 1962 Giacconi and Gursky successfully launched their rocket-mounted detector, spinning the rocket to allow the detector to scan the sky in all directions.

As they monitored the progress of the flight from the launch-site blockhouse, the crew in the control room were able to see readings from the detector via the rocket's radio telemetry. Almost immediately after the doors in the rocket opened, they saw a large peak in the X-ray count rate as the rocket spun past a point in the southern sky. Some of the crew were jubilant: they had succeeded in detecting the Moon! But Gursky wasn't so sure. The source was too bright. 'I knew what the rate should have been and I knew we would have to add all the data together before we had a chance to determine the signal accurately. So I felt we were in trouble,' he later recalled.

As they frantically processed the data, the team eliminated instrumental effects and the Moon as possible sources of the signal, which was actually coming from a position 30 degrees off from the Moon, in the constellation Scorpius. About 60 degrees from the main peak was another strong source of X-rays, located in the constellation Cygnus.

By late August, the AS&E group were confident enough to announce their discovery of these sources of cosmic X-rays, and, with Friedman's group, quickly confirmed their results in three rocket flights in 1962 and 1963. The X-ray sources were called Scorpius X-1 and Cygnus X-1, following the established convention of naming radio sources after the constellations in the sky in which they were situated. Because the early detectors and telescopes had very poor angular discrimination, few of the first X-ray sources that were discovered could be matched to specific known space objects within their general constellation areas. However, Friedman's group went on to identify the Crab Nebula as a celestial X-ray source, which became known as Taurus X-1, by flying a detector on a rocket to look at the Crab Nebula at the time that it was being covered by the Moon and watching its X-rays fading away. Taurus X-1 is a supernova remnant with a newly created neutron star in its centre. Scorpius X-1 eventually turned out to be a blue neutron star. Cygnus X-1 is a black hole.

With the advent of satellite observatories, which enabled astronomers to make observations that lasted much longer than the few minutes of a rocket flight, X-ray astronomy took a great leap forward. The first satellite entirely devoted to X-ray astronomy was Uhuru, a project led by Giacconi and named with the Swahili word for 'freedom', as it was launched from Kenya in 1970 on the twelfth anniversary of the nation's independence. Uhuru was a spinning spacecraft, able to survey the whole sky; it discovered 339 new sources of X-rays. Giacconi was eventually awarded the Nobel Prize in 2002 for his 'pioneering contributions to astrophysics, which have led to the discovery of cosmic X-ray sources'. Satellites like Uhuru and its successors enabled astronomers to identify many X-ray sources as binary stars. The X-rays are generated by the infall of matter from one star into the strong gravitational field of its companion, usually a neutron star or black hole. Scorpius X-1 and Cygnus X-1 are in this class. Other sources of X-rays are supernova remnants (in which the X-rays are from interstellar gas energized by the explosion), Seyfert galaxies and quasars (the X-rays are from supermassive black holes), clusters of galaxies (intergalactic gas energized by motions of the galaxies and their black holes) and gamma-ray bursters (X-rays derived from the explosion that makes the burst). Surely this gallery of extraordinary discoveries justified Rossi's declaration of faith: nature is infinitely more imaginative than man.

OUR GALAXY

AND ITS STARS

Variable and Eclipsing Stars
Discovery of star systems

As tho' a star, in inmost heaven set,
Ev'n while we gaze on it,
Should slowly round his orb, and slowly grow
To a full face, and there like a sun remain
Fix'd – then as slowly fade again,
And draw itself to what it was before.

Alfred, Lord Tennyson, 'Eleänore VI', 1832

For many centuries, stars were thought to be constant and unchanging. Astronomers were puzzled when they first noticed that some stars varied in brightness or even faded away entirely only to appear again later. Early Arab astronomers called one of these stars 'The Demon Star'. Although there is nothing demonic about their periodic disappearance, the modern explanation for this 'ghostly' behaviour is every bit as astonishing.

In 1596, while observing the planet Mercury, David Fabricius of Friesland in the Netherlands, a German pastor and a disciple of Tycho Brahe, noticed that a star that he had earlier used as a positional reference had inexplicably brightened and then faded away. At first, he believed it to be a nova, like the one that had been observed by Brahe in 1572, but the star then reappeared. Jan Fokkens Holwarda (sometimes called Johann Phocylides), also of Friesland, discovered in 1638 that Fabricius's star faded and came back every eleven months. In 1642, it was named Mira (Latin for 'wonderful') by Johannes Hevelius of Danzig; it is also called Omicron Ceti. Fabricius did not live to enjoy the recognition for his discovery, since he was murdered in 1617 by a peasant whom he had accused of stealing a goose.

In 1667 the Italian polymath and astronomer Geminiano Montanari noticed that the star Beta Persei also varied in brightness. Beta Persei is traditionally called 'Algol' (Arabic for 'The Ghoul', or 'The Demon Star'), which suggests that early Arab astronomers had observed these mysterious changes. Its variability was rediscovered in 1744 by a farmer and amateur astronomer, Johann Georg Palitzsch, who lived near Dresden, and was recorded again in 1782 by the English amateur astronomer John Goodricke.

Goodricke was born in 1764 in Gröningen, the son of a British diplomat and a Dutch woman. At five years of age he caught scarlet fever and became deaf. His parents had him educated at a special school for the deaf in Edinburgh; he learned to lip-read well enough to study at the Warrington Academy near York, where he became interested in astronomy. On 12 November 1782, when he was only eighteen years old, Goodricke recorded in his journal his discovery of the variability of Algol, which he observed was very regular. The star's brightness usually stayed near a magnitude of 2.1 but every 2.867 days it suddenly dropped to a minimum brightness of magnitude 3.3. Exactly halfway between the main minima there was a smaller dip in brightness. When Goodricke reported his observations in 1783 to the Royal Society of London, he offered two alternative explanations for the star's peculiar behaviour: that Algol was periodically occulted by another, dark body; or that Algol rotated and had a big spot on one side that made the star appear darker when the spot was facing Earth. We now know that the second theory does not apply to Algol, but it is a good explanation for the behaviour of other types of variable stars.

Goodricke's first theory was correct. Algol is in fact not one but two stars, one bright and one dim, each orbiting around the other. The orbit is almost edge-on to Earth, and when the larger, dimmer star completely covers the smaller, brighter star, it causes the most acute drops in brightness. When the smaller, brighter star passes in front of part of the larger, dimmer one, it produces the smaller dips that Goodricke had noticed.

The most puzzling feature of Algol's double-star system is that the less massive (dimmer, orange) star is more advanced in its evolution than the more massive (brighter, white) star. Usually it is the other way around: the more massive stars typically go through their life cycles more quickly than their less massive sister stars. The inexplicable reversal of these circumstances in Goodricke's star is known as the 'Algol paradox'.

The mystery was explained by American astronomer John Crawford in 1956. Crawford proposed that the more massive star had indeed evolved faster, in the usual way. But when this star expanded and became an orange giant, some of its material leaked onto the less massive, less evolved star close by, increasing its mass. Many stars are in binary systems and many are close enough for this exchange of material to happen, with outcomes that can be surprising.

It proved impossible to explain Mira in a similar way. There is no interposing body that causes its light periodically to dim: it is not a double star. The brightness of Mira cycles with a period of nearly a year as it pulsates, throbbing in size like a beating heart and, at the same time, changing in its temperature. The combination of the change in size and temperature changes the star's brightness. Cepheid variable stars are similar. The discovery of variable stars has helped astronomers to account for a variety of exotic specimens in the astronomical zoo.

Sirius B and White Dwarfs
Discovery of stellar cinders

> It is often stated that of all the theories proposed in
> this century, the silliest is quantum theory. In fact,
> some say that the only thing that quantum theory
> has going for it is that it is unquestionably correct.

Michio Kaku, *Hyperspace*, 1995

An astronomer's offhand remark to a colleague and a student's mathematical puzzle designed to pass the time during a long sea-journey led to the discovery of white dwarfs: dying stars so small and dense that they can throw other stars out of orbit, or implode into black holes.

As a young man, Friedrich Bessel worked as an accountant for a shipping company, where he developed interests in navigation and then astronomy. At the age of twenty-six he became the Director of the Königsberg Observatory in Prussia, and for the rest of his life he measured star positions with the observatory's telescopes. In 1844 Bessel noticed that the brightest visible star, Sirius, was progressing across the sky in a wavy motion. He realized that Sirius had an unseen companion star that was pulling it from side to side, disturbing its orbit. This was the first star to be discovered solely by means of its gravitational effect on another, although it was not actually seen until eighteen years later.

In 1862 the American telescope maker Alvan Clark was testing a new refracting telescope he had made for Dearborn Observatory in Evanston, Illinois, by inspecting the image that it formed of Sirius. His son, Alvan Graham Clark, Jr, saw the faint satellite star, which was almost lost in the white glare of Sirius. By chance, at the time of the Clarks' observation, the faint companion star, Sirius B, happened to be at the point in its orbit when it was furthest from its much brighter parent, and therefore easiest to see. The orbital period of Sirius B around Sirius A is fifty years.

In 1914, at the time of Sirius A's next large separation from B, the first spectrum of Sirius B was obtained by the Mount Wilson astronomer Walter Adams. Adams's spectroscopy showed that Sirius B was a little hotter than Sirius A, although ten thousand times less bright. Since Sirius A and B are both at the same distance from Earth, Sirius B has to be much smaller than Sirius A – less than 1% of its size, which is a shade smaller than the Earth and very tiny as stars go.

Sirius B is actually a white dwarf: the burnt cinder of a dying star. The first white dwarf ever identified was a star called 40 Eridani B, which was of similar brightness and temperature to Sirius B. William Herschel discovered 40 Eridani to be a double star (it is in fact a triple). In 1910, on a routine visit to Harvard, Princeton astronomer Henry Norris Russell pointed out to Harvard Observatory director Edward Pickering how faint 40 Eridani B was, and that it must be rather small, mentioning rather wistfully how useful it would be to know the star's temperature so that its size could be determined. Pickering happened to be directing a mass-photography project to find the temperatures of large numbers of stars. Russell's 'Eureka!' moment came as a mundane telephone call to Pickering's assistant Williamina Fleming. Russell later recalled that 'in half an hour she came up and said "I've got it here…" I knew enough, even then, to know what it meant….At that moment, Pickering, Mrs Fleming and I were the only people in the world who knew of white dwarfs.'

Russell produced in 1913 a diagram in which he organized star data in such a way as to make clear the extraordinary nature of 40 Eridani B. He plotted the brightness of nearby stars relative to their temperatures. Ejnar Hertzsprung had published a different version of the same diagram a few years earlier, hence its name: the 'Hertzsprung–Russell diagram'. Russell noticed a small, anomalously placed group of stars, one of them 40 Eridani B.

Its temperature was very high: it was 'white' hot. But it was also very dim, which meant that it was very small – a 'dwarf'. Russell correctly surmised that the star was a similar size to the Earth, although its mass was not unusual for a star in a binary system.

Approximately 95% of stars end their lives as white dwarfs. (Our Sun will.) A typical star becomes a red giant, then a planetary nebula and then a white dwarf, which passively fades away to a dark, dense stellar cinder. A white dwarf's mass is typically similar to the Sun's, but its size is much smaller, which makes it exceptionally dense – a matchbox filled with white dwarf material would weigh a tonne – and the force of gravity at its surface is very strong. The material inside a white dwarf star is extraordinarily strong as it has to withstand the tendency of the star to collapse under its own weight. In 1925 a young British physicist, Ralph Fowler, discovered that this material is 'degenerate': all the electrons are packed together as closely as is physically possible. The pressure generated by the degenerate material stops the star from collapsing.

Fowler's principles were applied to the structure of white dwarfs by a nineteen-year-old Indian mathematician, Subrahmanyan Chandrasekhar, who in 1930 was on his way from India in an ocean liner to study at Trinity College, Cambridge. Through the calculations he made to pass the time on the journey, Chandrasekhar discovered that there is a counter-intuitive relationship between the mass and the radius of a white dwarf – the more massive the star, the smaller its size. This means that there is a maximum mass above which a white dwarf cannot exist. This limit is known as the Chandrasekhar mass, and it is about 1.4 times the mass of the Sun. If a white dwarf gets more massive than this it shrinks to a point and becomes a black hole. As proposed in 1973 by the young British astronomer John Whelan and American theoretician Icko Iben, this is the scenario that creates some types of supernovae. Extra material dribbles onto a white dwarf from a nearby star, increasing its mass above the maximum, which causes it to collapse, release huge amounts of energy and finally explode.

However, when Chandrasekhar presented his results to his colleagues in 1935, he was publicly humiliated by the most distinguished astronomer in Britain at the time, Sir Arthur Stanley Eddington, who called the result 'stellar buffoonery'. In reaction to this incident, Chandrasekhar abandoned his intention to work in Britain and emigrated to the USA, where he worked at the University of Chicago for the rest of his life. Chandrasekhar was awarded the Nobel Prize in 1983 'for his theoretical studies of the physical processes of importance to the structure and evolution of the stars', in particular for the work on white dwarfs and black holes.

Walter Adams, director of the Mt Wilson Observatory, carried out work on other aspects of white dwarfs, of which Eddington was much more supportive. In 1925 Adams had discovered that light from the surface of Sirius B was redshifted. That is, light that set out from the surface of the white dwarf lost energy as it climbed out of the gravitational field of the star. As the light lost energy, it became redder. This was a predicted effect of General Relativity, and verified the high mass and small size of Sirius B. Elated, Eddington reported that 'Adams...has confirmed that matter 2000 times denser than platinum [degenerate matter] is not only possible but is actually present in the Universe.'

Neutron Stars and Pulsars
Stars that should not exist

> Oh, I'd rather they were neutron stars with rapid axial spin,
> And even pulsing white dwarfs would cause me no chagrin,
> But suppose they're radio beacons
> Guiding creatures with slobbery, malevolent grin?
> Oh, I'd rather that the pulsars had a natural origin.

An anonymous astronomer at the University of Michigan,
Ann Arbor, 1968

The discovery of pulsars was a serendipitous by-product of an investigation with completely different aims, which was intended to study the twinkling of radio stars. Then a young PhD student noticed an odd 'bit of scruff' on a data chart. Among the twinkling stars was a strange astronomical object that had never before been seen.

In the 1960s radio astronomer Antony Hewish and his colleagues in Cambridge built a radio telescope known (in splendidly archaic units) as the '4½-Acre Telescope' (the telescope's collecting area was 1.8 hectares). The telescope was intended to look at the scintillation, or twinkling, of radio 'stars'. The twinkling of ordinary stars is caused by irregularities in the Earth's atmosphere; the scintillation of radio stars is caused by irregularities in the plasma from the Sun that pervades the Solar System. Radio sources scintillate if they appear point-like, which is often the case if they are a long way away. Hewish's telescope was meant to identify twinkling radio sources that were quasars – exploding black holes – embedded in galaxies at vast distances. To see the twinkles, the telescope had to respond to changes in a radio source's intensity on a very short timescale, which required a large collecting area – hence the 4½-acre surface of wire netting that covered the stationary radio telescope. The enormous 'mirror' looked straight up and surveyed a strip of the sky as it rotated above the telescope.

By 1967 the telescope was ready and Hewish assigned a PhD student, Jocelyn Bell, to the job of analysing the data. She surveyed a 400-foot strip of chart paper for signs of the scintillating radio sources, rejecting terrestrial radio interference such as aircraft or TV stations. In October 1967 she noticed what she called 'a bit of scruff'. It was passing through the beam of the radio telescope in the middle of the night, when scintillation caused by the Sun was at a minimum, which pointed to it being terrestrial interference. However, Bell was not convinced: 'Sometimes within the record there were signals that I could not quite classify. They weren't either twinkling or manmade interference. I began to remember that I had seen this particular bit of scruff before...'

Bell and Hewish decided to use a faster recorder to get a clearer view. By November she had got a satisfactory recording which showed clearly that the 'scruff' was a burst of pulses almost exactly 1.5 seconds apart, similar to many kinds of terrestrial interference. When Bell told Hewish, he said 'Oh that settles it. It must be man-made.' Nothing of the sort had been seen in astrophysics before that varied so quickly and with such regularity.

However, as Bell studied the 'bit of scruff' further, it became obvious that it wasn't man-made. It stayed exactly in the same position in the sky, so it was celestial. It exhibited no signs of motion, so it was not in orbit around the Sun – it lay beyond the Solar System, among the stars. Bell had soon discovered three more sources of 'scruff', which she confirmed by backtracking through 3 miles of paper recordings. For a brief period, the Cambridge radio astronomers even wondered whether the sources of scruff were interplanetary craft, possibly navigation beacons, and jokingly numbered them LGM 1, 2 and so on (for Little Green Men). The lack of any orbital motion seemed to rule this out, since the sources were nowhere near any other star or sun. Finally, the true identity of the mysterious objects was announced, both in a sensational paper for the magazine *Nature* and as a dry appendix to Bell's thesis on the interplanetary scintillation of radio sources.

Bell had discovered the first examples of new astronomical objects called 'pulsars', a contraction of 'pulsing radio stars'. A pulsar is a small rotating neutron star. The pulsations arise because there is a kind of lighthouse-like rotating beam of radio waves on the neutron star that sweeps in the direction of the Earth once per rotation. The period of the pulsations is the rotation period – the star rotates about once per second. It can only do this because it is so tiny compared to other stars – a neutron star is roughly of radius 10 kilometres, so it would fit over a typical large city.

Bell's discovery actually confirmed an older theory that by the 1960s had been all but forgotten. Neutron stars had been theoretically predicted in the 1930s. In 1933 California astronomers Walter Baade and Fritz Zwicky had suggested that the release of gravitational potential energy as an ordinary star collapsed to a neutron star was the source of the energy of supernovae. In 1939 physicists Robert Oppenheimer and George Volkoff calculated the structure of a star made of neutrons, realizing that it was so compact that its gravity was governed by General Relativity and the pressure inside that held the star up was the repulsion that one neutron has for another, as if the star was a gigantic atomic nucleus. Oppenheimer and Volkoff thought at the time that nature had no way of actually making neutron stars and that their calculation was entirely theoretical. Their suggestions were all but forgotten until revived to explain pulsars.

In 1974 the discovery of pulsars was recognized by the award of a share of the Nobel Prize for Physics to Antony Hewish for 'pioneering research in radio astrophysics' for his 'decisive role in the discovery of pulsars'. The fact that Jocelyn Bell, as a student working under Hewish's supervision, was not awarded the prize jointly was a matter of controversy, with the provocative astronomer Fred Hoyle criticizing the circumstances. Hoyle wrote: 'Miss Bell's achievement...came from a willingness to contemplate as a serious possibility a phenomenon that all past experience suggested was impossible. I have to go back in my mind to the discovery of radio-activity by Henri Becquerel for a comparable example of a scientific bolt from the blue.' (In 1896 Becquerel was experimenting with a uranium-bearing crystal and left it in a drawer with photographic paper. When he opened the drawer some time later and developed the paper, the crystal had made its own photograph from radiation given off by the radioactive decay of the uranium.)

That Jocelyn Bell had been disregarded and the Nobel Prize for her discovery awarded only to her male supervisor remains to this day a feminist issue, but the affair changed the Nobel Prize Committee's rules – nowadays prizes for discoveries made during PhD studies are awarded to both the supervisor and the student. Bell herself has been honoured many times since then, and she has wryly remarked that not getting the Nobel Prize is better than getting one, because you get wonderful consolation prizes, including in her case being appointed a dame.

Black Holes
A solution looking for a problem

> Apollo to Mission Control –
> We are almost within reach of our goal,
> But our readings of g
> Seem excessive to me,
> So we may be inside a black ho...

G. J. S. Ross, 'Space Travel', 1975

The existence of black holes was predicted as early as the eighteenth century, but it was not until the 1970s that astronomers found some.

Imagine a star in space. A projectile is flung from its surface, like a ball thrown up from Earth. If the projectile is thrown at low speed, it rises from the surface and falls back. But there is a speed called the escape velocity, which is just fast enough to allow the projectile to escape from the star's gravity. The escape velocity of the star depends on its mass and its size – the more massive and smaller it is, the faster its escape velocity. A star with a combination of small size and large mass might have an escape velocity faster than the speed of light. Nothing can travel faster than the speed of light (according to the theory of relativity), so it would be impossible to throw a projectile into space from such a star. The projectile would always fall back. This is the basic idea underlying the concept of a black hole. It is a body in space – a planet, a star, or something similar – whose mass and size combine to give it an impossibly large escape velocity.

The Cambridge cleric and professor of geology John Michell discovered the concept of black holes in 1783. He speculated about the effect of gravity on light from the Sun. If the principle of escape velocity applied to light in the same way that it did to solid projectiles, the Sun's gravity would slow the flow of light out into space. For our modestly sized Sun the effect would be small, but Michell calculated that if the Sun was 500 times its actual size, so that its mass was 100 million times heavier, its gravity would be so strong that light would slow to a halt – it would not make it as far as the Earth and we would not be able to see the Sun. Alternatively, if the mass of the Sun shrunk into a sphere that was only 3 kilometres in diameter, it would generate the same effect. Pierre-Simon Laplace, a director of the Paris Observatory, put forward the same concept in 1795. However, since Relativity had not yet been discovered, Michell and Laplace did not know that the speed of light is constant, so these theoretical explanations for black holes were not entirely satisfactory.

Around 1910 the modern theory of black holes was expressed more correctly in terms of Albert Einstein's General Relativity by the German mathematical physicist Karl Schwarzschild. Schwarzschild put forward the following scenario: space curves around a massive body due to the gravitational distortion of spacetime, which causes light to follow curved paths (geodesics). If a body is sufficiently massive and sufficiently small, then light from the surface of the body might curve so tightly that it might reach no more than a small distance from the body. The body would then be black, because light would never leave it. The properties of bodies like this were summed up in the name given to them by the American physicist Robert Dicke around 1961 and popularized in 1967 by the theoretical physicist John Wheeler: 'black hole'.

The surface that divides a black hole from the outside world is called the event horizon. News about anything that happens inside the event horizon cannot escape the event horizon because light and other radiation that carries the news is dragged back by the strength of the black hole's gravity. Just outside the event horizon, gravity strongly bends the tracks of light rays from anything happening there, and its image is very distorted. The image of the event horizon of the black hole in the galaxy M87, the first to be photographed, is distorted in this way (plate XXVIII).

Although their theoretical basis had become quite advanced, black holes had never actually been observed in nature, so for a long time they were a solution looking for a problem. We now know that nature makes black holes in at least two ways: by supernova explosions in stars, and in the nuclei of active galaxies. Isolated black holes are dark and difficult to see. However, if matter (gas) falls into a black hole, it releases gravitational energy, which heats the gas. This can happen if the black hole has a companion star that leaks gas onto it, or if other stars get drawn near the black hole, break up and then fall into it. These two scenarios make some black holes visible as, on the one hand, X-ray binary stars and, on the other, active galactic nuclei.

X-ray binary stars are close binary stars that orbit around each other in short periods (ranging from minutes to days). One component is more-or-less normal star and the other is a much smaller star like a white dwarf, neutron star or a black hole. The normal star transfers matter (gas) onto its compact companion via an accretion disc: matter spirals inwards and falls onto the surface of the compact companion. The impact of the matter on the surface of the companion star makes the gas hot enough (a temperature of 10 million K) to radiate X-rays.

The orbit of the normal star depends on the mass of its companion, which in turn depends on what type of star it is. White dwarfs have a maximum mass of 1.5 times the mass of the Sun and neutron stars have a maximum mass of 3 times the mass of the Sun (in fact, no known neutron star has more than twice the Sun's mass). If the mass of the compact object exceeds 2 times the solar mass, it can only be a black hole. To apply this method in practice, astronomers must have a clear view of the normal companion in the binary system, and it must be a normal type of star so its mass can be estimated accurately.

In 1971 I was working with Louise Webster at the Royal Greenwich Observatory. The Observatory had developed a new spectrograph for the 2.5-metre Isaac Newton Telescope, and we were testing the instrument by measuring the motion of various stars, including the star called HDE 226868, which appeared in the same direction of the sky as Cygnus X-1. I suspected there might be a connection between them. HDE 226868 is a normal blue supergiant star and we did not think there was anything about it that would cause it to emit X-rays. However, we thought that its motion would change if it was circling around an X-ray-emitting companion. The first two or three spectra that we took were disappointing – there was no change of motion and we considered giving up. Then we found a spectrum that showed a large change. We had unlimited access to the telescope because we were testing the new instrument, so we continued, and with the next few spectra the cyclic change of motion became clear. HDE 226868 was moving around a companion. Later we realized that by unlucky coincidence we had taken the first few spectra at times when the star was at the same position in its orbit, which made it appear stationary. We also realized that the initial large change that had remotivated us was actually a false reading – we were using a new instrument and were not practised with it. Luckily, the error had inspired us to go on.

Because HDE 226868 is a supergiant, it has an unusual evolutionary history, so there is considerable uncertainty about its mass – it could be anywhere between 12 and 20 times the mass of the Sun. Nevertheless, we had discovered that the other X-ray emitting star in Cygnus X-1 is actually a black hole, with a mass greater than 4 solar masses (no less eminent a physicist than the late Stephen Hawking said that he was 95% certain that it was a black hole). Louise and I did not know that Tom Bolton, a Canadian astronomer working independently in Toronto, was following exactly the same train of thought as us and was coming to the same conclusion; he is also credited as a co-discoverer of the black hole in Cygnus X-1. Together we had found the problem for which Michell, Laplace and Schwarzschild had already provided the solution.

Of course, astronomy moved on and clearer examples of black holes were discovered, in particular through techniques involving the detection of gravitational waves.

Distances of the Stars

The radiance of that star was shining years ago

> Were a star quenched on high,
> For ages would its light,
> Still travelling downward from the sky,
> Shine on our mortal sight.

Henry Wadsworth Longfellow, 'Ode to Charles Sumner', 1875

When we look at distant stars in the night sky, we are actually looking into the past. Since antiquity, astronomers knew that if a star was observed regularly from the same position, it ought to appear to move very slightly each year. They sought to discover these small movements to measure the distances of the stars from Earth. Most stars are so far away that their light takes years to reach the Earth.

Efforts to calculate the distances of the stars from Earth date back more than 2,000 years. The Greek astronomer Aristarchus of Samos in the fourth century BCE and Copernicus in 1543 CE both realized that if the Earth moves around the Sun, the fixed stars should move in a reflection of the Earth's motion. This is manifestly not the case – the stars move very slightly, but nowhere near as much as the Sun's rising and setting. Both men came to the same conclusion: the stars are much further away than the Sun, and consequently the radius of the Earth's orbit is negligible compared with the distance of the stars.

The apparent movement of a star due to the Earth's motion around the Sun is called the star's annual parallax. 'Parallax' means the apparent shift of something due to the motion of the observer. Hold your finger up at arm's length, and keep it still, but move your head from side to side. The finger moves against the background. The angle by which it moves is its parallax.

In 1580 the Danish astronomer Tycho Brahe built a vast pre-telescopic sighting instrument, called the Great Mural Quadrant, at his Uraniborg observatory on the island of Hven to measure star positions. It was mounted on a wall built precisely north–south to measure the altitude of stars as they passed due south. It had a brass measuring scale with a 2-metre radius, and was at the time the most accurate instrument ever built to measure star positions, but even Brahe could not determine the parallaxes of stars, as they were so far away. He concluded that the stars were more than 700 times more distant than the Sun.

Dutch scientist Christiaan Huygens took a different approach. If the Sun is a star and all stars are the same brightness, then the reason why the stars are so much fainter than the Sun is that they are further away and their light is diminished by distance – in fact by the amount of their distance squared. Huygens thought that if you could measure the relative brightness of the Sun and a star such as Sirius, then by finding the square root of the difference you could calculate the relative distances of the Sun and the star. Huygens tried to measure the brightness of Sirius relative to the Sun by covering the Sun's face with a card pricked with different-sized holes. He matched the appearance of sunlight through the smaller holes with Sirius. His estimate, published posthumously in 1698, was that the distance of Sirius is 27,664 times the distance of the Sun.

This is a difficult measurement to make because the contrast in brightness between the Sun and Sirius is so great. The Scottish mathematician James Gregory tried a variation on Huygens's technique in 1668 by comparing the brightness of Sirius to that of a planet. He chose a time when a given planet was at its greatest distance from the Earth and roughly the same brightness as the star. He then waited until the planet was much nearer the Earth, and bright enough to compare with the Sun. He used his knowledge of the planet's distance to link together all his measurements. Using this approach, Gregory estimated the distance of Sirius at 83,190 times the distance to the Sun. Isaac Newton likewise calculated the distance to a typical bright star as 1 million times the distance of the Sun, but regarded this number (which was in fact much closer to modern estimates) as controversial and did not pursue the topic to further conclusions.

Unfortunately, this method for measuring the distances of the stars was fatally flawed: it incorrectly assumed that all stars are of the same brightness. With the rise of instrumental technology in the eighteenth and nineteenth centuries – telescopes that produced sharp images over large areas of the sky, and with finely calibrated scales for measuring angles – the geometrical method became the most promising technique. In 1669 Robert Hooke built a telescope on the side of a wall of his London house, intending to measure the parallax of Gamma Draconis, a star that passes directly overhead in London, but gave up after a few inconclusive measurements, illness and an accident that damaged the telescope lens. Samuel Molyneux and James Gregory tried again in 1725 and established that the parallax of Gamma Draconis was less than 1 arc second ($\frac{1}{3600}$ of a degree). On this basis they concluded only that the star was at a distance from the Earth further than 200,000 times the Sun's distance. They were right – the Gamma Draconis is in fact at 5 million times the Sun's distance.

Molyneux and Gregory's experiments made it clear that a nearby star needed to be selected as the target if the geometric method was to have any chance of success. But without knowing the relative distances of the stars beforehand, how would astronomers select a nearby star on which to try the measurement technique? Using a much larger telescope than was available to Molyneux and Gregory, but a similar technique, Wilhelm Struve, a German astronomer in Dorpat (now Tartu, Estonia), chose to measure the parallax of the bright star Vega in 1837 on the assumption that bright means close. Struve first measured a parallax of $\frac{1}{8}$ of an arc second, but got $\frac{1}{4}$ of an arc second when he repeated the experiment. This discredited his methods. A Scottish astronomer, Thomas Henderson, working in South Africa in 1832–33, measured the parallax of the star Alpha Centauri and would have got a reasonably accurate result, but did not analyse his data until he returned to Britain, and even then could not quite believe what he had found and kept reworking his analysis.

The German astronomer Friedrich Bessel had a better result when he chose the star 61 Cygni because it tracked quickly across the sky. This indicated that it was close to Earth, much as a close-passing bee will move more quickly across your field of vision than a high-flying aircraft. He also used a new technique. Essentially, he measured the angle between 61 Cygni and a star that was very nearby, and watched how this angle varied throughout the year. Bessel's measurement of 61 Cygni in 1838 from the observatory in Königsberg, Prussia (now Kaliningrad, Russia), with a parallax of $1/8$ of an arc second was believed. Bessel's work boosted Henderson's confidence in his own result for Alpha Centauri and he published his measurement in 1839. Historians rightly credit Bessel as the astronomer who discovered the true distances of the stars.

The most accurate stellar parallaxes have been measured from space by the Hipparcos satellite, led by its project scientist Mike Perryman of the European Space Agency. Hipparcos was an acronym for 'High precision parallax collecting satellite', with a deliberate echo of the name of the ancient Greek astronomer Hipparchus of Nicaea. From 1989 to 1993 the satellite precisely measured the distances of 120,000 stars and of 2 million stars to a lower degree of accuracy. The data generated by this satellite will be superseded in the 2020s by the Gaia satellite, launched in 2013, which is gathering data on 1 billion stars.

As we measure the distances to the stars, we actually study the past. Expressed in terms of the time that light takes to travel, the distance of the Sun is 8 light minutes, but the distance of 61 Cygni is 9 light years. That is to say, the light that glimmers down at us from 61 Cygni is actually the light that was emitted by the star nine years ago. The lights that we see when we look up at the night sky originated at many different times long ago, and some show the memory of stars that have since reached the end of their lives and no longer exist. Our view of the sky is thus a view that is peculiar to us, a mosaic of various distances and epochs that meld into a single view for us here and now. We see the stars in Orion (2,000 light years distant) as they were when Christ walked in Jerusalem. To us they are new-born stars; but to anyone in the Andromeda galaxy (2.5 million light years away), these stars haven't formed yet. On the other hand, there are stars in Andromeda that have already exploded and died in their own galaxy, but still shine on us in ours.

The Discovery of our Galaxy
Stars in an island universe

> Our present evidence, so far as it goes, leads to the belief that
> the spirals are composed of great clouds of stars so infinitely
> distant that we cannot make out the individual stars.

H. D. Curtis, *The Nebulae*, 1917

The discovery of nebulae in the seventeenth century opened the
doors onto the rest of the Universe. Viewed through the lenses of the
newly invented telescopes, the Milky Way resolved into individual
stars and astronomers realized that we were not alone. Beyond our
Sun there were other suns, and beyond our Galaxy, other galaxies.

In 1609 and 1610 Galileo turned his telescope on the night
sky, and was amazed when he looked at the Milky Way. The milky
luminescence, visible to the naked eye and thought at the time to
be a seam, a road or a cloud, split into thousands of stars. Galileo
announced his discovery in his 1610 treatise *Sidereus Nuncius*
('Starry Messenger'): 'We are at last free from earthly debates
about the nature of the Milky Way. It is, in fact, nothing but a collection of innumerable stars grouped together in clusters. Upon
whatever part of it the telescope was directed, a vast cloud of stars
is immediately presented to view. Many of the clouds are rather
large and quite bright, while the number of smaller ones is quite
beyond calculation.' Excited by Galileo's report, other astronomers
followed him in inspecting the sky, chancing on interesting clouds
or 'nebulae'. The Bavarian mathematician Simon Marius discovered the Andromeda Nebula in 1612, describing it as looking like
a 'flame seen through horn' (lanterns in Marius's day had windows
made from thin sheets of horn) as the nebula had a characteristic
elliptical shape. The Andromeda Nebula is just visible to the naked

eye and had actually been noted by the Persian astronomer 'Abd al-Rahman al-Sufi in 964. In pre-Islamic cultures the constellation of Andromeda was depicted as a fish. At the fish's mouth lay a fuzzy patch, the first known representation of the Andromeda Nebula, which, in his *Book of Fixed Stars*, Al-Sufi called 'the Little Cloud'. It was rediscovered again in 1654 by the Sicilian priest Giovanni Battista Hodierna, who compiled a list of up to forty similar nebulae.

The list of 'nebulous stars' grew longer as astronomers started observing with telescopes systematically, measuring the positions of all the stars in the sky and noting what they looked like. The astronomer Johannes Hevelius from Danzig measured 1,564 stars, listing sixteen as nebulous in his posthumous catalogue of 1690. John Flamsteed, the first British Astronomer Royal, catalogued 2,935 stars and mentioned several that were nebulae; his successor Edmond Halley added six more, including the 'star' Omega Centauri that he observed from Saint Helena, where it was possible to see more of the southern part of the sky. In an expedition to the Cape of Good Hope, South Africa, to survey the whole of the southern sky, Abbé Nicolas-Louis de Lacaille discovered dozens of nebulae while he was measuring the positions of uncharted stars and inventing new southern constellations.

Between 1764 and 1781 the comet-hunter Charles Messier compiled all these discoveries, along with some of his own, into a catalogue of nebulae that eventually totalled over one hundred entries. The catalogue became an essential tool for astronomers, a source list of objects on which to test new theories and instruments.

Messier's catalogue was sent to William Herschel for review on the same day that he was elected to the Royal Society for his discovery of Uranus, which inspired him to examine all the entries with his new telescopes. He then began to search for and classify other fainter nebulae. His method was to sweep the telescope over the sky in a raster pattern (a scanning pattern of parallel lines looking like a rake), noting double stars and nebulae from his perch high on the telescope, shouting down the details to his sister Caroline, who took notes at a table below. Herschel described this process (in eighteenth-century spelling) as 'star gaging'. He had expected to add only a few nebulae to Messier's list, but, in the end, he found more than two thousand. With his telescope, Herschel was able to resolve many of the nebulae he found into individual stars, and, like Galileo, thought that all would eventually succumb to his increasingly powerful instruments.

Meanwhile, in 1750, the English mathematics teacher Thomas Wright had developed his theory of the Milky Way. He modelled it as a band of light that arises from a flattened slab of stars. Look from within the slab, in the plane, and you see many stars and much starlight; look across the slab and you see fewer stars, and thus less starlight. An account of Wright's theory was published in a newspaper in Königsberg where it was read by the philosopher Immanuel Kant, who was inspired to work on the problem. He proposed in 1755 that the Milky Way star system was flattened because it was rotating, and was held together by its own gravity. Kant also proposed that other nebulous patches like the Andromeda Nebula were similarly rotating masses of stars like the Milky Way, held together in the same way.

The implications of Kant's second suggestion were astounding. Although by the eighteenth century astronomers were comfortable with Galileo's revelation that the Earth was not the centre of the Solar System, and some even had speculated that there might be more than one planetary system, the idea that there could be more than one galaxy required another mighty transformation of perspective. Before they could convince the public, however, astronomers needed to prove that the theory was correct.

Herschel gave quantitative expression to Kant's suggestions as he investigated the structure of the Milky Way by counting the density of its stars. He thought that the number of visible stars in a given area of the sky would indicate the extent of the Milky Way in that direction. Using this technique, Herschel discovered that the stars filled a flattened circular structure that he compared to the shape of a grindstone. He suggested that the Andromeda Nebula is another 'Milky Way' – a galaxy like our own, seen at an angle; a closely compressed cluster of stars that would eventually be individually resolved from the nebulosity. As time went on, he became less sure of this, and made contradictory statements; when he died in 1822, no one was really sure what Herschel thought he had discovered.

Herschel became confused because in his day the classification of 'nebulae' was complicated. Originally the word meant something that looked amorphous, like a cloud. Some nebulae, such as the Orion Nebula, are indeed true clouds of gas that can never be resolved into stars, although stars may be embedded within the clouds. Modern astronomers still call these gaseous objects 'nebulae', but other cloudy patches, originally called nebulae, are in fact clusters of stars, which may be densely packed together or distant. The weakness of the instrumentation at the astronomer's disposal or the enormous distance of the cluster prevents this second type of 'nebula' from being resolved into individual stars.

At the beginning of the twentieth century, the development of larger telescopes, like the 24-inch telescope at the Boyden Observatory in Arequipa, Peru, and the 60- and 100-inch telescopes at the observatory on Mount Wilson in California, allowed astronomers to see individual stars in some of the closer 'nebulae' when the night air was still and clear. Some were variable stars, Cepheids, whose distances could be determined, paving the way for American astronomer Edwin Hubble's dramatic 1925 discovery that some 'nebulae', like the Andromeda Nebula, are indeed distant galaxies of stars like our own Milky Way, separated from it. Our Galaxy is, in the picturesque phrase of the day, an 'island universe', one of many.

Interstellar Nebulae
Stars, molecules, dust and gas

An unformed fiery mist, the chaotic material of future suns.

William Herschel on the Orion Nebula, 1789

When William Herschel viewed the nebula in the Sword of Orion through his telescope, he described it as a 'fiery mist' that looked like the flame of a candle (plate XXII). What he saw is actually a small dent on the side of a Giant Molecular Cloud, most of it dark. Its shadowy fields of dust and gas conceal thousands of infant stars and new-born planets.

Nicolas-Claude Fabri de Peiresc was a French aristocrat who trained as a lawyer and became a diplomat. He was also an amateur scientist. In 1610, when a friend acquired a telescope similar to Galileo's, Peiresc used it to discover a nebula (now known as the Orion Nebula) surrounding a star in the Sword of Orion. Galileo had already discovered three of the four small stars in the little cluster (now known as the Trapezium) that illuminates the nebula, but had not noticed the nebulosity around them. In 1656 the nebula was rediscovered by the Dutch scientist Christiaan Huygens, who published the first sketch of it. The nebula was recorded as number 42 in Charles Messier's catalogue.

On receiving a copy of Messier's catalogue, William Herschel turned his telescope to observe the Orion Nebula, which he described as 'an unformed fiery mist, the chaotic material of future suns'. Herschel thought that he had detected changes in the overall shape and brightness of the nebula over the decades that he observed it. This claim was controversial, but potentially very significant, as it suggested that the nebula was not a large, distant object, but small and relatively close to the Earth. If the nebula was very large, its individual parts would not be able to communicate with each other quickly enough for their brightnesses to change at the same time. Nor could the nebula alter quickly if it was made of independent stars.

Herschel's son, John Herschel, made a special point of mapping the nebula very carefully in 1826 so that any changes could be proved in future observations. A decade later he remapped the Orion Nebula and found that his father had been mistaken about the changes: he saw no differences between the new map and the one he had made ten years earlier: John Herschel's findings were confirmed when the first photographs of the Orion Nebula were made independently in 1880–83 by two pioneers of astronomical photography, the American amateur astronomer Henry Draper, and the English retired businessman Andrew Ainslee Common, who photographed the nebula to test a new technique. Draper made the first photograph of the Orion Nebula in 1880 in a 50-minute exposure. Common was an engineer and used his engineering skill to build a telescope that could track a nebula smoothly and accurately enough for its image to be recorded on the photographic emulsions available in the 1880s, which required long exposure times. His observatory in Ealing near London contained a 36-inch reflecting telescope. In 1885 it was bought by Edward Crossley, a member of the British parliament, and given by him to Lick Observatory ten years later, where it is still in use under the name of 'the Crossley Reflector'. Common wrote about his photograph: 'Although some of the finer details are lost in the enlargement, sufficient remains to show we are approaching a time when photography will give us the means of recording, in its inimitable way, the shape of a nebula and the relative brightness of the different parts in a better manner than the most careful drawing.' Draper's and Common's images proved that the brightness of the Orion Nebula was unchanging; it is large and far away.

Because the Orion Nebula is so bright, it has always been a natural target for people testing new technology. In 1864 William Huggins deployed his new astronomical spectroscope on the brightest nebulae. Instead of the continuous rainbow spectrum of colours that would indicate that the light from the nebula was similar to sunlight, and therefore made by stars, Huggins saw the individual spectral lines of a glowing gas. Astronomers had speculated that all misty 'nebulae' would eventually be found to be a mass of stars like the Andromeda galaxy or the Pleiades. Huggins's result showed that this was not true – some nebulae, like the Orion Nebula, were truly gaseous, although they may contain a few individual stars.

Huggins was especially puzzled by a strong green line in the nebula's spectrum. Between 1880 and 1889, he and his wife Margaret persistently tried to photograph the spectrum, and discovered that the strong green line did not have a wavelength that coincided with any known terrestrial element. The Hugginses incorrectly attributed the line to a new cosmic element, which they called 'nebulium'. The light represented by the mysterious line gives the nebula a green colour when it is viewed by the naked eye, although photography or the more modern electronic detectors show the nebula as reddish, because the technological processes emphasize the red light generated by burning hydrogen more than the human eye does.

The Orion Nebula contains not only gas, but also copious amounts of dust. One especially dusty region shows as a protrusion of dark material into a bright nebula near to the Orion Nebula that is illuminated by the star σ (sigma) Orionis. The dark nebula and its distinctive horse-head shape were discovered in 1888 by Williamina Fleming. She had emigrated to the USA in 1878 from her native Scotland, and after her husband abandoned her with a young child, supported herself by working as a maid in the home of Edward Pickering, the director of the Harvard College Observatory. Eventually Pickering asked her to take on clerical work at the observatory, and then promoted her to scientific assistant. The story is that Pickering exhorted his male assistants to achieve more with the admonition 'My maid could do better!' and then found that, indeed, she could. Fleming discovered the Horsehead Nebula as a dark indentation recorded on a photograph taken by Edward's brother, William Pickering.

The Orion Nebula and the Horsehead Nebula are just two features in a vast complex of dust, gas and stars that eventually became known as the Orion Giant Molecular Cloud. It is 1,500 light years across and covers the entire constellation of Orion. The bright object that Herschel and the Hugginses had seen is only a hollow dent on the surface of the enormous cloud, like the white flesh of an apple left after the first bite from its red surface. The inner surface of the dent on the surface of Orion's Cloud is illuminated by the four Trapezium stars.

The concept of 'giant molecular clouds' (GMC) was proposed in the 1970s by radio astronomers who mapped the radio emission from molecules inside the Orion cloud. The radio emission was first detected in 1963, when Sander Weinreb of the National Radio Astronomy Observatory at Greenbank, West Virginia, and MIT physicist and engineer Alan Barrett identified the presence of the hydroxyl molecule (OH) in the cloud. Over the next few years, ammonia (NH_3), water (H_2O), formaldehyde (H_2CO) and carbon monoxide (CO) were also detected. All these chemicals are commonly found on Earth, where they would normally be broken down by light from the Sun, but in the Orion GMC the massive dark clouds had shielded the molecules from the destructive effects of starlight, allowing them to survive in large quantities. These molecules also play a key part in the evolution of the cloud, and end up in the by-products of the formation of stars – comets and planets – as 'seeds' that develop, it seems, into the chemical building blocks of life.

How do astronomers know that there are stars inside the dark cloud? The dust of the Orion GMC can be very dense and hides the light of most of the stars inside it. Infrared radiation can penetrate the dust, and following the development of the technology in the mid 1960s, astronomers have been able to detect infrared radiation from the hidden stars. Pioneers like Eric Becklin, Gerry Neugebauer,

Frank Low and Douglas Kleinmann discovered individual sources of infrared by laboriously scanning the cloud with single detectors in raster scans, just like the survey of the sky that William Herschel made with his conventional telescope.

These early infrared readings were constantly hampered by the abundant infrared radiation emanating from the Earth itself, but in the 1980s the US military declassified a range of infrared detectors that had been developed during the Cold War to identify the warm engines of approaching vehicles and rockets, and which were capable of taking pictures. Astronomers eagerly exploited their ability to see warm objects in the sky, using them not only in ground-based telescopes but also in satellites, such as IRAS (InfraRed Astronomy Satellite, launched 1983), ISO (Infrared Space Observatory, 1995), SIRTF (the Space Infrared Telescope Facility, later renamed the Spitzer Space Telescope, 2003) and the Herschel Space Observatory (2009). The satellite-mounted detectors operate in the cool of space, carefully avoiding interference from the Earth's and Sun's infrared radiation.

These developments have made it possible to see that the Orion Giant Molecular Cloud, and others like it, contain thousands of stars, which recently formed inside the cloud. Approximately two thirds of these stars are orbited by planet-forming discs – new Solar Systems in their infancy. Herschel's prescient description of the Orion Nebula as 'the chaotic material of future suns' could be extended in modern times to 'the chaotic material of future suns, planets and life itself'.

Star Clusters
Nebulae resolved

> Many a night I saw the Pleiads, rising thro' the mellow shade,
> Glitter like a swarm of fire-flies tangled in a silver braid.

Alfred, Lord Tennyson, 'Locksley Hall', 1842

With the invention of the telescope, certain individual stars and nebulae that had been known since antiquity were revealed in fact to be dense clusters of stars. As astronomers mapped these clusters and elucidated their properties, they began to comprehend the astounding size of our Galaxy and to understand how stars age and, eventually, die.

Some star clusters have been known since antiquity. The Pleiades and the Hyades are readily visible in the constellation of Taurus and are mentioned in the *Iliad* and the Old Testament. The Praesepe star cluster, known as 'the Beehive', was described by the third-century BCE Greek poet Aratos and by the second-century BCE astronomer Hipparchus. When Galileo turned his telescope on the sky in 1609 he looked at the Praesepe and discovered that it was a star cluster: 'The nebula called Praesepe is not one star alone, it is a mass of more than forty small stars.'

Praesepe, the Pleiades and the Hyades are all star clusters in our Galaxy, loose conglomerations of perhaps a few hundred or a few thousand stars. Other star clusters are more densely packed and spherical, containing hundreds of thousands or even millions of stars, orbiting around and through our Galaxy. These dense, mobile clusters are called 'globular clusters'. The discovery of the first globular cluster was reportedly made in 1665 by the obscure German amateur astronomer Johann Abraham Ihle. In 1677, during an expedition to Saint Helena, Edmond Halley discovered one of the nearest and largest globular clusters, best visible from the

Southern Hemisphere and known as Omega Centauri (plate XIII). This designation takes the form of the name of a star, which is how it appears to the naked eye, perhaps a slightly fuzzy one. Describing it as a 'lucid spot or cloud', Halley could not resolve the cluster into individual stars with his telescope.

William and Caroline Herschel found many 'nebulae' during their 'sweeps' of the sky, and with their superior telescopes identified a proportion of the 'nebulae' as star clusters. William classified their shapes, labelling some of them as 'globular'. His son John Herschel noticed that the globular clusters were not uniformly distributed across the sky, but were concentrated towards the constellation Sagittarius. In 1909 the Swedish astronomer Karl Bohlin deduced from this that the globular clusters surround the centre of the Galaxy, which also lies in the direction of Sagittarius. The American astronomer Harlow Shapley used Henrietta Leavitt's method for determining the distance of variable stars in the clusters to map their three-dimensional distribution. In 1918 he estimated the distance to the centre of the globular cluster system as some 60,000 light years, and the diameter of the whole system of globular clusters as 300,000 light years. Although these values are twice the currently accepted estimates, Shapley had discovered that our Galaxy is astoundingly large – much larger than Galileo and the Herschels had ever imagined.

The origin of the globular clusters is enigmatic. They seem to be a mixture of star clusters that are native to our own Galaxy and others that have fallen into our Galaxy when it absorbed galaxies that it encountered at random in intergalactic space.

By contrast to the neat spherical look of the globular clusters that orbit high above the plane of our Galaxy, straggly, irregular-looking star clusters congregate in the plane of the Galaxy near its spiral arms. The early observers could not know that these clusters would be the keys that unlocked the mystery of how stars evolve. Star clusters allow astronomers to develop techniques for comparing different stars, because they eliminate many of the usual uncertainties and differences from star to star. All the stars in a given cluster are at the same distance from the Earth, so the light of each is dimmed to the same extent by distance. All the stars were formed at the same time from a single gas cloud. They are of the same age and composition, but have different masses.

On the basis of these assumptions, the Danish astronomer Ejnar Hertzsprung was able to discover between 1907 and 1911 that the brightnesses of most of the stars in a cluster correlate to their temperatures. He found that in both the Pleiades and the Hyades the brighter stars were hotter (blue dwarfs) and the fainter ones were cooler (red dwarfs). Henry Norris Russell noticed the same correspondence in nearby stars. It transpired that the hot, bright stars were the more massive ones and the cool, dim stars less massive. In old clusters the bright blue stars were missing, and there were bright red stars instead (red giants and supergiants): the blue stars became bright red stars as they aged. In even older clusters many of the red giants had died and become white dwarfs, neutron stars or black holes

Putting all the different clusters in order of age, the situation became clear. The bright, massive stars aged quicker than the dim, less massive ones, becoming red giants and then dying faster than the others. Massive stars have more hydrogen fuel than the less massive ones, but they burn it much faster, and consequently run out of energy and die earlier, just as a profligate millionaire may go bankrupt more quickly than a poor miser.

The internal structures of the different kinds of stars were first calculated by Arthur Stanley Eddington and James Jeans between 1916 and 1924. In the 1950s and 1960s, with the advent of the first electronic computers, it became possible to link these calculations together to track how stars changed from one kind to another as they aged. The results could then be used to date stars in different star clusters, each cluster consisting of an array of stars of all possible masses, but each star having reached a different stage in their life history.

For decades, star clusters were the main way of verifying theoretical calculations about the size, composition and life cycles of stars. There was no way to see past the surface of a star in order to study what was really happening inside it. This changed in the 1970s with the discovery of helioseismology and solar neutrinos, which proved that these calculations had been remarkably accurate. Star clusters were the ancient keys to a modern problem, keys with which astronomers were first able to unlock the opaque outer layers of the stars in order to discover their secrets.

Supernovae

Origins of the stardust from which we are made

> All these stupendous objects are daily around us; but because
> they are constantly exposed to our view, they never affect
> our minds, so natural is it for us to admire new, rather
> than grand objects. Therefore, the vast multitude of stars
> which diversify the beauty of this immense body does not
> call the people together; but when any change happens
> therein, the eyes of all are fixed upon the heavens.

Saint Basil the Great (Bishop of Caesarea), fourth century CE

Before the sixteenth century, stars were thought to be fixed and
eternal. We now know that stars are in constant flux, undergoing
a cycle of birth, death and rebirth. The roadside observations of a
sixteenth-century nobleman and the striking pictures beamed to
earth by the Hubble Space Telescope since 1990 bear vivid witness
to supernovae, the explosive collapses of dying stars. This process
generates the basic elements that make up all of the matter in the
universe, including the building blocks of our own bodies.

One winter's evening in 1572, the Danish nobleman Tycho Brahe
was returning home in his carriage when he was struck by something
unusual in the night sky. It looked like a new star. Brahe halted his
carriage and asked passing peasants to confirm what he had seen:

> On the 11th day of November in the evening after sunset,
> I was contemplating the stars in a clear sky. I noticed that a
> new and unusual star, surpassing the other stars in brilliancy,
> was shining almost directly above my head; and since I had,
> from boyhood, known all the stars of the heavens perfectly,
> it was quite evident to me that there had never been any star

in that place of the sky, even the smallest, to say nothing of a star so conspicuous and bright as this. I was so astonished at this sight that I was not ashamed to doubt the trustworthiness of my own eyes. But when I observed that others, on having the place pointed out to them, could see that there was really a star there, I had no further doubts.

The 'new star', or 'nova' (short for *nova stella*), had in fact been noticed as early as 6 November 1572 by others, including Francesco Maurolico, a star-gazing Sicilian mathematician, and Hieronymus Muñoz, a Spanish philosopher, who saw the nova while giving an open-air evening class. Astronomers Michael Mästlin of Tübingen, Thomas Digges of Kent and Brahe himself all measured the position of the nova by noting that it lay on the intersection of the lines joining certain pairs of stars. They showed that the new star did not move through space (as a comet would), nor did its position in the sky change when the Earth's rotation changed the position of the person observing the new star. This proved that the nova was a long way from Earth, among the fixed stars. Brahe wrote: 'I conclude, therefore, that this star is not some kind of comet or a fiery meteor, whether these be generated below the Moon or above the Moon, but that it is a star shining in the firmament itself – one that has never been seen since the beginning of the world.'

This momentous discovery challenged the ancient Christian and classical concept that the stars were eternal and unchanging. In the cosmology of philosophers such as Aristotle, the orbits of the Moon and Sun marked the boundary between the changeable and permanent parts of the universe. The Moon had phases, the Sun had spots, but the stars were the same forever. The Earth was imperfect and harboured misery, disease and sin, but the stars were perfect and pure, the home in heaven for the blessed saints. Brahe's new star was irrefutable evidence that this worldview was wrong.

The star was in fact a supernova. This word was coined by the Swiss-American astronomer Fritz Zwicky in 1931 when he discovered that there were some new stars that were much brighter than others. These new stars were releasing colossal amounts of energy, their brightness making the surrounding galaxy appear dim in comparison.

A supernova does not represent the birth of a new star, but the destruction of an old one. Stars are in constant balance between two opposing forces: the downward force of gravity and the upward force of pressure generated by the heat and density within the star. Nothing turns off the downward force of gravity. However, in ordinary stars, the amount of upward pressure depends on the energy generated by nuclear processes within the star. Stars contain a finite amount of nuclear fuel, and eventually the fuel gives out. The exhausted star may stabilize as a white dwarf, but some stars are too massive to stabilize. If the internal pressure can no longer support the structure of the massive star, the star collapses and releases a lot of energy in freefall. It is this energy that causes the star to explode and become visible as a 'new star' – where the old star had been too dim to be noticed.

The explosion causes the old star to disintegrate (plate XVIII), although usually not completely – it may leave a black hole or a neutron star. The body of the star, including all the elements made inside it by nuclear fusion, is dispersed into space as it disintegrates. Eventually, this material mixes with interstellar gas, and congeals to form new stars and planets. This happened in our own Solar System before our present Sun was formed, and elements from ancient supernovae constitute the physical makeup of our own planet and everything on it – including our own bodies. We are made of stardust.

The supernova explosions just described are called Type II supernovae. There is an alternative kind. Brahe's supernova was a so-called Type Ia supernova – its progenitor star was a white dwarf star. White dwarfs are held up against the force of gravity by a force of quantum mechanical origin called degeneracy. As discovered by Subrahmanyan Chandrasekhar, this balance only works if the white dwarf is not too massive. If the white dwarf feeds on material donated from a close companion star, its mass increases and may surpass the upper limit. The star collapses in a supernova explosion, in which the white dwarf is completely disrupted. Its companion star, the one that donated the extra material, is released like a stone from a slingshot. The companion star released by Brahe's supernova was discovered in 2004 by a team led by Spanish astronomer Maria Pilar Ruiz-Lapuente. She used the William Herschel Telescope on La Palma in the Canary Islands to identify the star, and later confirmed her discovery by observing the star more closely with the Hubble Space Telescope. 'Here we have identified a clear path: the feeding star is similar to our Sun, slightly more aged,' Ruiz-Lapuente reported. 'The high speed of the star called our attention to it.'

In the case of Type II supernovae, the progenitor stars are actually too massive to support themselves from the outset. Massive stars are ticking time-bombs, and all of them eventually collapse and become supernovae. Because stars are long-lived, in a galaxy the size of the Milky Way there is on average only one supernova every fifty years. But this does mean that if you watch fifty galaxies, you will find one supernova exploding every year, and if you watch a thousand galaxies, you will find a couple of dozen every week. This approach has made it possible for the Hubble Space Telescope to find and study many faint supernovae at great distances. Astronomers working on the Supernova Cosmology Project and the High-Z Supernova Search Team use large ground-based telescopes and powerful automatic image analysis systems to search fields crowded with galaxies that lie just ahead in the path of the HST. When the teams locate supernovae in these galaxies, they can direct the HST to study them in detail, discovering more of the secrets of these explosive dying stars.

Supernova 1987A
The whisper and the vision

So when, by various turns of the Celestial Dance,
In many thousand years
A star, so long unknown, appears,
Though Heaven itself more beauteous by it grow,
It troubles and alarms the World below.

Abraham Cowley, 'Ode to the Royal Society', 1667

Most supernovae are in distant galaxies and therefore rather faint. The appearance of a bright, nearby supernova like Tycho Brahe's 'new star' is always an occasion for excitement, and a source of valuable astronomical information. The most recent supernova that could be seen on Earth with the unaided eye was called SN 1987A (indicating that it was the first supernova of 1987). In 2005 astronomers studying the supernova witnessed an astoundingly beautiful and unprecedented event.

SN 1987A was discovered by Ian Shelton at 05:40 GMT on 24 February 1987 at Las Campanas Observatory in Chile. He had taken a photograph of the Large Magellanic Cloud and developed it before going to bed. To his surprise there was a black spot on the photograph, which, at first, he thought was some sort of blemish. Then he realized that it was actually a bright star where none was indicated on the charts. It was a supernova.

Shelton wanted to share his discovery, and went to another telescope to talk to his colleagues. A fellow astronomer, Oscar Duhalde, mentioned that he had seen the star earlier that night with his own eyes while strolling about outside during a break, but when he returned to the telescope he had been immediately harassed by another researcher in the queue for the equipment and had forgotten to mention the strange object. Later it transpired that the supernova had in fact been photographed the day before by astronomer Robert McNaught in Australia, but he had put off looking at his photographs and thus failed to discover the nova before Shelton.

Shelton's discovery made it possible to identify the first neutrino particles from a supernova. When a Type II supernova collapses, its protons and electrons are jammed so close together that they merge and form neutrons. This nuclear reaction produces huge numbers of elusive particles called neutrinos. Some of these neutrinos were caught in a neutrino detector in Gifu, Japan, called Kamiokande (Kamioka Nucleon Decay Experiment). In its latest version, called Super-Kamiokande, it is a cylindrical tank, 40 metres tall and 40 metres in diameter, filled with 50,000 tonnes of purified water. Hemispherical photomultiplier tubes lining its inside wall catch flashes of light produced when neutrinos are absorbed by the water. The main use of Kamiokande was to catch neutrinos originating from the Sun. Upon learning of Shelton's discovery, the Japanese scientists searched their computer files and discovered an unusual burst of neutrinos at 07:35 GMT on 23 February, the day before the explosion had made the nova bright enough to be noticed. An American team operating a neutrino detector in a salt mine in Ohio then searched their records, and reported that their detector had recorded a burst of eight neutrinos at the same time – although these were mere whispers compared with the enormous scale of the explosion. The neutrinos from SN 1987A were the first from outside the Solar System to be discovered, providing remarkable insight into the conditions in the hitherto unseen interior of a supernova explosion.

Neutrinos are by-products of the nuclear reactions that build up the heavier elements from hydrogen and helium inside stars. One of the main types of nuclei made in a Type II supernova is called nickel-56, which eventually becomes cobalt-56 and finally iron-56 through the process of radioactive decay, emitting gamma rays in the process. At first these gamma rays are absorbed by the expanding body of the exploding star. But eventually the explosion thins out and the gamma rays can escape. Gamma rays from SN 1987A were discovered by a satellite-mounted gamma-ray detector called Solar Max, which had been designed to study the Sun. It was only by chance that the satellite was in orbit at the time the supernova went off and that its design let it detect gamma rays from a completely different celestial object.

As SN 1987A faded away it became easier to probe into the neighbourhood of the star, and astronomers realized that something else about it was interesting. In 1989 Joe Wampler discovered that the spectral lines of a small nebula were visible in the supernova's spectrum. The spectral lines were confirmed by an orbiting satellite called IUE. The next year the Hubble Space Telescope imaged a ring that surrounded the supernova, and, in 1994, discovered that the ring was in fact a symmetrical, hollow, three-dimensional bipolar structure, shaped like two glass tumblers set bottom to bottom. This small nebula had been produced by the star that exploded as SN 1987A during previous phases of its life, some 20,000 years before the explosion witnessed by Shelton. By 1998 the HST was able to see that the exploding supernova was about to crash into this smaller nebula. By 2005 the whole of the central ring was involved in the crash, and the nebula lit up like a celestial firework display. One problem that SN 1987A threw up was that even thirty years after the event, no neutron star has appeared as the material from the supernova dissipated, as had been predicted. Possibly the supernova made a neutron star and then, after material immediately fell back onto it, the neutron star collapsed further to a black hole.

Cepheid Variable Stars
Beats of a star's heart that measure the Universe

As tho' a star, in inmost heaven set,
Ev'n while we gaze on it,
Should slowly round his orb, and slowly grow
To a full face, and there like a sun remain
Fix'd – then as slowly fade again,
And draw itself to what it was before.

Alfred, Lord Tennyson, 'Eleänore VI', 1832

Cepheid variable stars have a regular 'heartbeat'. Thanks to the work of pioneering female astronomer Henrietta Leavitt, this stellar heartbeat makes it possible to measure the size of the Universe.

In 1784 English astronomer John Goodricke discovered that the star Delta Cephei was a variable star. It pulsates, expanding and contracting like a heart beating, the regular changes in size causing its brightness to increase and fade between magnitude 3.6 and 4.4. Goodricke estimated the period of its brightness cycle as 128 hours and 45 minutes: 5.36634 days.

Delta Cephei became the prototype for explaining similar pulsating variable stars, which were subsequently given the generic name of 'Cepheids'; other examples have periods that range from a few days to several hundred days. These stars are heat engines: they convert heat energy that they make in their interior into mechanical motion that moves their outer layers up and down. In the 1930s the English astronomer Arthur Stanley Eddington explained how this worked in theory; the details were confirmed by the Russian astronomer S. A. Zhevakin in the 1950s. Essentially, there is a valve mechanism in the star. The valve is closed when the star is small, causing pressure to build up and the star to expand. When the valve opens, heat and pressure are allowed to escape and the star shrinks back to its initial size.

The 'valve' is a layer of ionized helium in the upper layers of the star. When the star is smallest, the helium is opaque and traps radiation. As the radiation increases, it lifts the ionized layer, so that the helium recombines and becomes transparent, releasing the radiation and causing the outer surface of the star to fall again.

Cepheid variable stars are fascinating in themselves, but they are especially important in astronomy because they are 'standard candles' that can be used to measure the distance of other galaxies. Normally the brightness of a star is only a rough indicator of its distance, as stars' brightnesses vary according to their sizes. But the special qualities of Cepheid stars make it possible to estimate their intrinsic brightness more precisely. One of the main projects of the Hubble Space Telescope, led by American astronomer Wendy Freedman, was to discover and measure the brightness of Cepheid variable stars in external galaxies in order to calculate their distances from the Earth and from each other. The ultimate goal of Freedman's team was to measure the size of the Universe. Henrietta Leavitt's discovery made this possible.

Henrietta Leavitt was born in Lancaster, Massachusetts, in 1868 and died too early, of cancer, in 1921. She studied at the Society for the Collegiate Instruction of Women (later Radcliffe College and now part of Harvard University) and in her final year took an astronomy course, which enthused her about the subject. An illness left her deaf, but, after she had recovered, she became a research assistant at Harvard College Observatory at a salary of 30 cents an hour, working for its director Edward Pickering. It was not a time when women in the field were encouraged to carry out independent research, although Leavitt was undoubtedly of the highest intellect and capability. In her work for Pickering she discovered 2,400 new variable stars, doubling the number then known. Some 1,800 of them were found on photographs of the Magellanic Clouds taken at the Boyden Observatory then at Arequipa, Peru. Some proved to be Cepheid variables.

In 1908 Leavitt discovered that the brighter Cepheids in the Magellanic Clouds took longer to complete their brightness cycles. Leavitt did not know that the Clouds were galaxies separate from the Milky Way, but she did reason that all the Cepheids in each Cloud were at the same distance from Earth, so their longer periods were not an illusion caused by distance but must somehow relate to their average light output. Leavitt's discovery became known as the period-luminosity relation, and demonstrated that Cepheids were 'standard candles' that could be used to measure distances by comparing the apparent and intrinsic brightnesses of recognizable stars. Good standard candles are consistent (neighbouring stars of roughly similar brightness), highly luminous (for observation at great distance) and reliably recognizable. Cepheids are easy to pick out in the sky because of their variability. Once you have determined the period of a Cepheid, this Cepheid is the same average brightness as others of the same period. You need to anchor the relationship somehow. This has in the past been done by locating Cepheids in close clusters of stars whose distance you can find by other methods, but recent space experiments have been making it possible to triangulate directly to some Cepheid stars.

After her untimely death, Leavitt's work made it possible for modern astronomers to measure the size of the Universe and discover the location of objects within it. Astronomers Ejnar Hertzsprung and Harlow Shapley found that the Magellanic Clouds were outside the Milky Way galaxy, at distances currently estimated as 160,000 light years from Earth. The Andromeda galaxy – the nearest galaxy of the same size as ours – is 2.5 million light years from Earth. Wendy Freedman's team has used Hubble Telescope observations of Cepheid variable stars to measure the distance of thirty-one galaxies, out to distances of 70 million light years.

The distance calibrators for more distant galaxies can be referenced against this scale. However, although the accuracy of the Cepheid distance scale has been vastly improved over the past century, there remain worrying discrepancies in the scale of the Universe when the Cepheid distance scale is compared to the distance scale measured in other ways. The discovery of gravitational waves from merging black holes offers the exciting prospect of being able to help astronomers with the distance scale, because the distance of the mergers comes naturally out of the observations. But it replaces one problem with another: identifying which out of millions of galaxies the merger is taking place in, so as to determine its redshift to relate to the distance. The argument about the expansion scale of the Universe is not yet over!

Exoplanets

Other worlds beyond ours

> For there is a single general space, a single vast immensity
> which we may freely call Void; in it are innumerable
> globes like this one on which we live and grow.

Giordano Bruno, *On the Infinite Universe and Worlds*, 1584

Until a few years ago astronomers knew of one planetary system in the Universe – our own Solar System. This changed in the 1990s, with the discovery of the first of a number of large, Jupiter-sized planets orbiting central stars. Yet none of these exoplanetary systems seems to have formed in the same way as our own Solar System.

For three thousand years philosophers had been theoretically convinced that there were other worlds like our own in existence. More recently, astronomers had realized that planetary systems are a necessary consequence of the formation of stars, and had detected proto-planetary discs orbiting young stars, for instance, in the Hubble Space Telescope images of the Orion Nebula. By the last decade of the twentieth century it was, in fact, becoming worrying that no actual planets had been found orbiting other stars.

Finally, in 1992, Aleksander Wolszczan, a Polish-born American radio astronomer, was timing the rapidly rotating pulsar PSR 1257+12 and he noticed that its pulses alternately arrived earlier and later than expected. The pulsar was being pulled nearer to and further from Earth by three earth- and moon-sized planets in orbit around it. Wolszczan had discovered the first planetary system other than the Solar System. But Wolszczan's system exists in circumstances completely different from our own: it is the remains of a supernova explosion in a binary star, and is a second-generation planetary system, formed not at the birth of a star but at its demise.

The system's central star was as unlike the Sun as it is possible to get.

In 1995 two teams of astronomers discovered three planetary systems orbiting stars that were much like our own Sun. Two Geneva Observatory astronomers, Michel Mayor and Didier Queloz, made the first discovery. In April 1994 they had embarked on a programme to detect any radial velocity variations in a list of 142 nearby Sun-like stars which could be due to the gravitational pull of jupiters (large gas planets). The analogy of our own Solar System guided the search. Although we typically say that the planets orbit around the central Sun, this is actually a simplification. In fact, the Sun and the planets orbit around their common centre of mass. Because the Sun is so much more massive than any planets, the Solar System's centre of mass is actually inside it, so the Sun scarcely moves in its orbit – its motion of 13 metres per second is not much faster than world sprinting records and is more of a slow quiver than an orbit. This delicate motion of the Sun is mostly caused by the gravitational effect of the most massive planet in the Solar System, Jupiter. The main period of the Sun's motion is therefore the same as the orbital period of Jupiter, namely twelve years.

To find signs of this 'quivering' motion in other stars, Mayor and Queloz developed a spectrograph capable of detecting such small oscillations. They studied the 142 bright, nearby solar-type stars that had been selected because they showed, at coarser accuracy, no large velocity changes that would suggest they were members of double-star systems. It seems that planets survive only in orbit around single stars: astronomers calculate that a planet in a double-star system would loop in complicated orbits among the two stars, and in a relatively short time would be ejected from the system.

Despite their careful planning, Mayor and Queloz had not expected to be successful so soon. Their original programme essentially decided their careers, since they could only be sure that they had found a twin of Jupiter by observing at least two periods – amounting to twenty-four years! But within only eighteen months they had discovered their first planet, orbiting the star 51 Pegasi,

45 light years distant. Even more surprisingly, the star's oscillation was much greater than expected, and its period was very much shorter – 4.2293 days! The planet, 51 Peg b, is roughly the same mass as Jupiter. Its orbit is close to circular and the orbital period of four days indicates that it lies much closer to its sun than does Jupiter to our Sun – in fact, closer than any planet in our Solar System. The distance of 51 Peg b to its parent star is only ½₀ of the Earth–Sun distance.

As news of the discovery spread around the world, Mayor and Queloz's observations were swiftly confirmed during a brief four-day observing run with the powerful Lick Observatory telescope conducted by Geoffrey Marcy of San Francisco University, Paul Butler of the University of California, and a team from the High Altitude Observatory and the Harvard-Smithsonian Center for Astrophysics. Like Mayor and Queloz, Marcy and Butler had been monitoring solar-type stars for radial velocity variations indicating the presence of jupiters, but 51 Pegasi was not on their original observing list because of a mistake in the catalogue from which they compiled their programme. After hearing of the Swiss discovery and altering their expectations accordingly, the team quickly found further examples of exoplanets, some of them in archives of previous observations that they had not yet examined closely because they thought there was no rush.

The first technique used to discover exoplanets uses spectros-copy, a technique that gives an important property of the exoplanet, namely an estimate of its mass. It is not a method, however, that readily lends itself to the discovery of lots of planets, because you can examine only one star at a time. A second technique that has been used to discover exoplanets is to look for 'winking stars': stars whose light is periodically dimmed a little by the transit of planets across their faces. The dark planet obscures a small fraction, typi-cally less than 1%, of the bright star. Since astronomers can image thousands of stars in one picture, they can examine many stars for the periodic winks – the hard part is to measure the brightness of the stars accurately enough and to seek the significant, periodic dips in brightness. This technique results in a complementary property of the exoplanets that it finds, namely their size. Furthermore, by scrutinizing the spectrum of the star for the effects of the transiting exoplanet, it is possible to determine properties of the atmosphere of the planet. For example, some exoplanets have expanding atmospheres, boiling off as a result of heat from the parent star.

The transiting technique has been used by exoplanet-hunters to find planets. Some of them are ground-based, like SuperWASP (Wide-Angle Search for Planets), an international project that uses two robotic observatories located in the Canary Islands and South Africa. This programme uses two arrays of eight wide-angle cameras to image the sky repeatedly as it passes overhead, and has found about 200 exoplanets so far. But because the accuracy with which astronomers can measure the brightness of stars from the ground is limited by the trembling of the atmosphere, the most productive method to discover exoplanets is from space. There have been two space observatories devoted to this method, the pioneering French CoRoT (*Convection, Rotation et Transits planétaires*) spacecraft, active between 2006 and 2012, and the incredibly productive NASA Kepler mission, which between 2009 and 2018 stared at a field of stars in Cygnus and detected more than 5,000 possible planets, with about 3,000 planets so far confirmed.

A third technique to discover exoplanets is to image them, but this is very difficult because exoplanets are faint and are close to a very bright star. The technique has been successful in few cases (plate XVII).

Due to technical constraints, only very specific kinds of exoplanets have been discovered thus far, representing a minuscule proportion of the total. Virtually all of the stars that are currently known to have exoplanetary systems are no more than 3,000 light years away – and therefore relatively bright; you need a lot of light to measure accurately the small changes in a star produced by a planet. Most of the known exoplanetary systems contain large planets orbiting a central star, because only the largest planets can be discovered – Neptune-sized or bigger. 'Jupiters' are the most commonly found. Often the jupiter in the extrasolar planetary systems is in an orbit much closer to its sun, and thus hotter, than the Jupiter in our Solar System. Extrapolating from the sample that we have, it seems that there are as many planets in the Galaxy as stars – roughly half the stars have no planet, roughly half the stars have an average of two planets each. It seems that the most common planets in the Galaxy as a whole are so-called 'super-earths', twice the mass of our own Earth. It is not known why we do not have a super-earth in our Solar System.

These new systems of planets, large and close to their central stars, contradicted astronomers' theories about the formation of our own Solar System. Large planets are not supposed to be so close to their suns, but are expected to orbit the colder outer reaches of a planetary system, like Jupiter, Saturn, Uranus and Neptune, which retain their gases because they are massive and cold. Astronomers had believed that planets as near to their suns as the ones discovered recently would be small terrestrial planets, resembling Mercury. The reason seems to be that new-born planets migrate in towards their sun, the bigger ones swallowing up smaller inner ones. But something happened in our own Solar System to stop or even reverse the migration of Jupiter. This is good for us: otherwise our Earth would not have survived.

The Energy of the Sun and Stars
Discovery of nuclear fusion

> That evening after we had finished our essay, I went for a walk
> with a pretty girl. As soon as it grew dark, the stars came out,
> one after another, in all their splendour. 'Don't they sparkle
> beautifully?' cried my companion. But I simply stuck out
> my chest and said proudly: 'I've known since yesterday why
> it is they sparkle.' She didn't seem the least moved by this
> statement. Perhaps she didn't believe it. At that moment,
> probably, she felt no interest in the matter whatever.

Fritz Houtermans, quoted in Robert Jungk,
Brighter than a Thousand Suns, 1958

How does the Sun shine, and how long has it been shining? Could
the fuel ever run out? These questions have perplexed scientists
since they first began to comprehend the awesome size and age of
the Solar System. The secrets of the Sun's power system – for better
or for worse – have made nuclear technology possible on Earth.

As soon as the scale of the Solar System and the distances of
the stars became apparent, it was clear that the amount of heat and
light emitted by the Sun and stars was enormous. When astronomers
calculated the mass of the Sun using Newton's laws of gravity, it,
too, was enormous. If the Sun is a kind of normal fire, providing
power by chemical means (for example, by chemically combining
carbon and oxygen to carbon dioxide, as happens when wood or
coal is burnt), there is a lot of fuel available to supply the power
that we see. But for how long could the fuel last?

Presumably the Sun is at least as old as the Earth – the one
depends on the other. In 1650, to determine the age of the Earth,
Archbishop James Ussher published an analysis of the chronology

of events in the Bible, which was reproduced in the standard edition of the Bible used in England for centuries and so became widely accepted. He set the date of the creation of the Earth at 4004 BCE. If the Sun really had been created along with the Earth just 6,000 years ago, and chemical energy was the source of its power, the amount of material consumed as the Sun shone was only a small fraction of its total mass. But in the nineteenth century, British geologists like Charles Lyell and John Phillips began to estimate that the Earth was actually millions of years old, based on calculations of how long it would take for sedimentary rocks to be laid down from sea deposits or for rocks to be eroded away.

This created a problem for the idea that chemical energy was the source of the Sun's power – a conventional fire or chemical reaction would not be able to continue burning for millions of years. The discrepancy was made worse when calculations by physicist Lord Kelvin and biologist Thomas Huxley suggested that the Earth was actually hundreds of millions of years old. Even this seemed short, considering the length of time needed for the evolutionary processes that Charles Darwin envisaged in 1859 to produce the variety of living species on Earth. The German physicist Hermann von Helmholtz and the Canadian astronomer Simon Newcomb temporarily rescued the chemical energy theory at the end of the nineteenth century by suggesting that the Sun's energy was also supplied by gravitational contraction, calculating that the Sun's lifetime could be consistent with the Earth's if they were both at least hundreds of millions of years old. But in the twentieth century scientists began to realize that the Earth was much older even than this.

After Henri Becquerel and Marie and Pierre Curie discovered radioactivity, the British physicist Lord Rutherford developed a technique for using radioactive decay to measure the age of rocks, by which a young American chemist, Bertram B. Boltwood, discovered that some rocks were as much as 1–2 billion years old. How could the Sun keep shining for such a length of time?

The answer was discovered by two physicists of the University of Göttingen, Fritz Houtermans and Robert d'Escourt Atkinson. As they passed their summer holiday in 1927 on a walking tour, they discussed the problem of the source of the Sun's energy. They knew about the physical conditions inside the Sun from the work of the British astrophysicist Arthur Stanley Eddington: a high density and a high temperature created a high pressure inside the Sun, which countered the force of gravity that was drawing the material of its body tightly together. Atkinson knew of Einstein's formula for converting mass to energy, $E = mc^2$, and understood that what was then called atomic transmutation was possible. The atoms (or, as we now know, their broken-down nuclei) in the centres of stars and the Sun are frequently colliding together because of the high densities and high temperatures. If the collisions transformed some atoms from one kind to another, losing mass in the process, atomic (or 'nuclear') energy would be produced. 'This might be the source of the Sun's energy. 'Let's just work the thing out, shall we?' said Houtermans. 'How could it happen in the Sun?'

The two young scientists discovered how the fusion of light elements into heavier ones could fuel the Sun. Atkinson later learned that the Sun was mainly hydrogen, and realized that the source of the energy was specifically the conversion of four hydrogen atoms to one helium atom. A helium atom is 0.7% lighter than four hydrogen atoms. This tiny amount of excess mass, multiplied many times over by the astonishing number of hydrogen atoms present in the Sun and the frequency of their collisions, provides the Sun's energy. 400 million tonnes of mass disappears from the Sun every second and is transformed into solar energy – this attrition rate can be kept up for billions of years.

Houtermans and Atkinson's work was followed up in 1939 by the German-American physicist Hans Bethe. He discovered the exact process that enables hydrogen fusion in the Sun. It is called the CNO cycle, since the hydrogen nuclei are fused together in successive stages with carbon, nitrogen and oxygen as intermediate steps. Bethe was awarded the Nobel Prize in 1967 'for his contributions to the theory of nuclear reactions, especially his discoveries concerning the energy production in stars'. The details of what happens inside the Sun have been confirmed with astonishing accuracy by the detection of neutrinos, tiny particles given out in the nuclear processes, which travel from the Sun's interior and have been detected on Earth using specially built neutrino detectors.

In the decades after Bethe's discoveries, physicists from many nations were organized into programmes that attempted to reproduce on Earth what happens in the stars, finding ways to release nuclear energy slowly in a reactor or suddenly in a bomb. For better or for worse, many of these projects have been successful. The discovery of the energy source of the Sun and other stars may well prove to have been the most momentous secret of the universe to be uncovered.

The Origin of the Elements
Making star stuff

> We are bits of stellar matter that got cold by
> accident, bits of a star gone wrong.

Sir Arthur Stanley Eddington, *The New York Times Magazine*,
9 October 1932

'All the innumerable substances which occur on Earth – shoes,
ships, sealing-wax, cabbages, kings, carpenters, walruses, oysters,
everything we can think of – can be analysed into their constituent
atoms,' wrote James Jeans. 'It might be thought that a quite incredible number of different kinds of atoms would emerge from the
rich variety of substances we find on Earth. Actually, the number
is quite small. The same atoms turn up again and again, and the
great variety of substances we find on Earth result, not from any
great variety of atoms entering into their composition, but from
the great variety of ways in which a few types of atoms can be
combined.' The remarkable fact is that all but one of the elements
in the Universe – and therefore the elements that constitute the
building blocks of our own bodies – are made inside stars.

In the eighteenth and nineteenth centuries, chemists realized
that all materials were made of molecules, and molecules themselves were made of atoms in fixed arrangements. In the twentieth
century the actual physical makeup of atoms was discovered. Each
atom is made of electrons orbiting a nucleus, itself made of protons
and neutrons. The number of electrons in an atom is equal to the
number of protons in its nucleus. The number of neutrons in an
atom is roughly equal to the number of protons, but can differ
from atom to atom of the same element. Each different nuclear
arrangement is called an 'isotope'. There are about a hundred

chemical elements, and the atoms of each are distinguished from the others by the number of electrons. Changes in the arrangement of the electrons produce light; astronomers can see the light with a spectroscope, and, in general, the clearer the spectral signature of a particular element in a celestial body like a star, the more of that element is present in the star – its 'abundance'.

In her 1925 doctoral thesis, Harvard astronomer Cecilia Payne (later Payne-Gaposchkin) suggested that hydrogen was the most abundant element in the Sun. Reviewing her thesis, the influential Princeton astronomer Henry Norris Russell dismissed the idea, but changed his mind in 1929. In fact 71% of the Sun's mass is hydrogen, 27.1% is helium, together comprising 99.9% of the number of atoms in the Sun. The remaining 0.1% are (in order of abundance) oxygen, carbon, nitrogen, magnesium, silicon and neon; some seventy further elements have also been identified in the solar spectrum. Payne-Gaposchkin had discovered that the abundance of elements in the stars was broadly the same as in the Sun, so astronomers could treat the Sun as representative of other stars.

After Fritz Houtermans and Robert d'Escourt Atkinson discovered in 1927 that the stars generate energy by nuclear reactions, astrophysicists were able to address the question of where the elements came from and why some are more abundant than others. In 1939 Hans Bethe showed how hydrogen was transformed into helium via a cycle involving carbon, nitrogen and oxygen. This explained one source of helium – but where did the hydrogen, carbon, nitrogen and oxygen come from in the first place? According to a 1948 paper known as 'αβγ' (alpha-beta-gamma) after its authors Ralph Alpher, Hans Bethe and George Gamow, the hydrogen was made in the Big Bang and elements were built up in stages from the simplest element, hydrogen, by successively adding neutrons one at a time to make heavier and heavier nuclei, including additional helium.

The 'αβγ' theory failed because it could not create elements heavier than lithium. The problem is that there is no stable atom with 8 protons and 8 neutrons, so when you get to 8 neutrons the nucleus decays spontaneously back to 7. Armagh Observatory director Ernst Öpik and astrophysicist Edwin Salpeter found a better explanation in 1951–2: carbon, with 12 protons and 12 neutrons, is made when three helium nuclei (each with 4 protons and 4 neutrons) collide at the same time inside a star. In 1953 British cosmologist and nuclear physicist Fred Hoyle discovered the exact nuclear reactions involved, although the bold predictions he made to tie up the loose ends in his theory were thought by many to be crazy. But he persuaded physicist Ward Whaling in the Kellogg Radiation Laboratory at the California Institute of Technology to perform an experiment, which confirmed his calculations.

Hoyle was the first to establish the currently accepted explanation for how the elements are made inside stars. Although some said that Hoyle's motivation was to find an alternative to the idea that the elements originated in the Big Bang, he had started work on the topic long before he proposed the rival steady-state theory. In studying the way that stars evolve he had simply asked himself the question, 'What would be the very last of the nuclear reactions that take place in stars, instead of the first reactions that had so far occupied the attention of astronomers?'

The detailed scheme by which the elements were made in stars was summarized by Margaret Burbidge, Geoffrey Burbidge, William Fowler and Fred Hoyle in 1957. Their seminal paper is known by astronomers as B^2FH ('B-squared FH'), from the authors' initials, and is one of the most frequently referenced papers in astronomy. The Burbidges provided the astronomical expertise for the paper and Fowler and Hoyle the nuclear physics. In 1983 Fowler received the Nobel Prize 'for his theoretical and experimental studies of the nuclear reactions of importance in the formation of the chemical elements in the Universe'; it is not clear why the Nobel Prize committee did not also honour Hoyle.

The B^2FH paper proposed that stars make helium by 'burning' hydrogen, which is the most simple and abundant element in the

Universe. The burning converts helium to carbon, carbon converts helium to oxygen, oxygen and subsequent burning produces neon, silicon and finally iron. When a star explodes as a supernova the explosion irradiates elements such as carbon and oxygen in the body of the star. Other elements are produced on the surfaces of red-giant stars – dramatic proof of this was provided in 1952 by Mount Wilson astronomer Paul Merrill, who discovered the spectral lines of the element technetium in some red giants. Technetium is radioactive and its longest-lived isotope decays relatively quickly – in a matter of a million years. Since red giants are much older than this, the technetium must have been made inside them.

The elements that are made in stars are dispersed into interstellar space by supernova explosions, stellar winds and planetary nebulae. There, they mix with hydrogen gas and form clouds that may eventually condense into stars. This is the origin of all the chemical elements that make up the Earth and all that is on it, including ourselves. In the words of Carl Sagan, 'We are star stuff.'

Inside the Sun
Whispers and rings

> There's not the smallest orb which thou behold'st
> But in his motion like an angel sings.

William Shakespeare, *The Merchant of Venice, c.* 1599

We cannot look at the interior of the Sun directly, but over the past century we have developed ingenious ways to 'see' what happens inside it. Astronomers first captured neutrinos, incredibly small particles generated inside the Sun, by burying a vast quantity of dry-cleaning fluid in a gold mine. Studies of solar earthquakes showed that the Sun rings like a gigantic bell and provided clues to its makeup.

Astronomers who wanted to understand what happens inside the Sun faced one big problem: none of its internal workings could be seen, because the Sun is completely opaque. At first only the Sun's surface characteristics, and its global properties – such as its diameter and the amount of energy that it radiates – could be determined by direct observation. However, we now know what happens inside the Sun thanks to three lines of astronomical enquiry. First, ingenious mathematical calculations built up a theoretical picture of the Sun's interior. This picture was verified by enormous efforts to capture a tiny quantity of the neutrino particles, and later by measuring the sound waves generated by motions inside the Sun.

Our understanding of the inner workings of the Sun is the result of one of the great feats of modern mathematical reasoning. From the 1920s astronomers knew the physical conditions inside the Sun by calculation and from the 1930s they knew that nuclear reactions were the source of the Sun's energy. In the 1950s they had begun to understand the way that stars evolve in relation to one another from observations of star clusters. These calculations built up confidence in astronomers' theoretical knowledge of the Sun's interior.

The physics suggested that the Sun would be an abundant source of neutrinos. Neutrinos are small nuclear particles, whose existence was suggested by the Austrian-Swiss physicist Wolfgang Pauli in 1930 to explain some details of nuclear reactions, and confirmed experimentally in 1956. Neutrinos are made in the nuclear chain reaction inside the Sun that makes the Sun's radiation from hydrogen. The nuclei of the hydrogen atoms in the Sun become free protons. Two protons combine, one of them changing to a neutron by emitting a neutrino and a particle called a positron. The reaction continues to its conclusion when another proton sticks to the pair, forming a helium nucleus, containing a pair of protons and a neutron. Two similar helium nuclei collide and two protons are ejected, leaving behind a helium nucleus with a pair of protons and a pair of neutrons.

The net result of this chain is that four protons make a helium nucleus, releasing energy. The neutrinos escape, also carrying off small parcels of energy. The numbers of neutrinos given off by the Sun is immense – about ten billion pass through every square centimetre of the Earth every second. There are floods of them, but they are whisper-quiet and can travel through a light year (ten trillion kilometres) of material without interacting with it in any detectable way. Neutrinos travel so fast that it takes only eight minutes for them to reach the Earth. Despite the astonishing speed and elusiveness of solar neutrinos, it is possible to build detectors that do catch some.

The first solar-neutrino detector was built by Brookhaven National Laboratory physicist Raymond Davis, Jr, following technical suggestions from the notorious Italian-born physicist Bruno Pontecorvo (who later defected to join the Soviet nuclear programme) and American physicist Luis Walter Alvarez. Three American nuclear astrophysicists – William Fowler, Alistair Cameron and John Bahcall – had insisted that it was practical to try to catch neutrinos, as the vast numbers constantly released by the Sun overwhelmed the small chance for each one that it would slip through a neutrino detector unnoticed. In the bowels of the Homestake Gold Mine, in Lead, South Dakota, deep enough underground to avoid interference from cosmic rays, Davis installed a tank containing 615 tonnes of carbon tetrachloride, a solvent normally used for dry cleaning. Solar neutrinos were captured on the chlorine atoms in the solution and converted to argon atoms. These argon atoms were flushed out of the tank every two months and counted.

The original estimates were that just seventeen argon atoms would be produced in the tank in each extraction run, but in fact, in the first experiment in 1968, lasting six months, even fewer neutrinos were seen. As Davis repeated his experiment with improved equipment, the question became 'where are the missing neutrinos?' – this became known as the 'solar neutrino problem'. Another neutrino detector called Kamiokande, built and operated by Japanese astrophysicist Masatoshi Koshiba, was able to determine the trajectory of the incoming neutrinos. It not only confirmed in 1989 that Davis had detected neutrinos from the Sun and that there were fewer than expected, but was able to prove that the neutrinos it captured really came from the Sun. The Sudbury Neutrino Observatory (SNO), a neutrino observatory located 2,100 metres underground in Vale's Creighton Mine in Ontario, Canada, detected solar neutrinos through their interactions with a large tank of heavy water – about 300 million dollars' worth, loaned by Atomic Energy of Canada. It likewise found that solar neutrinos were missing.

At first, some physicists thought that the discrepancy between observation and theory had arisen because astronomers' standard calculations relating to the solar interior must be flawed. There were no missing neutrinos, but somehow astronomers had overestimated the numbers of neutrinos that the Sun was making. The astronomers rejected this, in part because they had found another way to look inside the Sun, to check their theories about its makeup and to solve the mystery of the missing neutrinos. The approach they used was called 'helioseismology'. Helioseismology is the study of oscillations in the body of the Sun, which resemble earthquakes studied by seismologists on Earth. In the general turmoil of motion of hot material in the Sun's interior, the Sun generates sound waves whose resonances travel across the body of the Sun, and its surface oscillates up and down. The Sun rings, like a bell quietly singing as it is brushed by a succession of impacts from a stream of sand grains.

Caltech physicist Robert Leighton discovered the surface oscillations of the Sun in 1960, and measured the oscillation periods at about five minutes. In the 1970s UCLA physicist Roger Ulrich suggested that the duration, frequency and tone of these oscillations could provide clues to the composition of the Sun's interior. Ulrich pointed out that the frequencies at which the Sun rings depend on the time it takes sound to cross the Sun. This in turn depends on the composition, temperature and density structure of the solar interior. The sound waves thus carry information about the interior of the Sun to the surface where it can be seen, just as the oscillations of earthquakes carry information about the interior structure of the Earth.

Individual earth-based telescopes had a great limitation: they could not observe the Sun after it disappeared daily below the horizon at night. Astronomers therefore set up networks of ground-based solar telescopes around the world to measure the frequencies of solar oscillations more accurately – the networks have names like GONG (Global Oscillation Network Group of the US National Solar Observatory), BiSON (Birmingham Solar Oscillations Network) and HiDHN (High Degree Helioseismology Network) – but intermittent cloud still interfered with their observations.

The Solar and Heliospheric Observatory (SOHO) satellite, a joint project involving the European Space Agency and NASA, avoided even this limitation. It has been staring at the Sun continuously from space since its launch in 1995. The comprehensive observations of the SOHO satellite provided new data on the temperature inside the Sun, and the way that its interior rotates slower than its surface layers, generating a hot layer inside the Sun that is the ultimate cause of sunspots and prominences on its surface. SOHO also proved that the standard calculations that had been used to measure sound-speed at various depths in the interior of the Sun were 99.9% accurate. The conclusion was that astronomers knew rather well how many neutrinos the Sun was making, and Davis's 'missing neutrinos' were not the result of a miscalculation of solar conditions.

Assuming that astrophysicists knew about the state of material inside the Sun and nuclear physicists knew how many neutrinos that would create, physicists had to concentrate on why many went missing. Something evidently happened to neutrinos after they had left the Sun. Some of them did not make it across space to the Earth. This explanation was first proposed by Pontecorvo only a year after Davis first found the solar neutrino discrepancy in 1968.

Neutrinos come in three different kinds, or 'flavours'. They can oscillate from one 'flavour' to another as they travel for eight minutes across the distance between the Sun and the Earth. The neutrino detectors have the capacity to capture and detect solar neutrinos only of the 'flavour' generated deep within the Sun. By the time the neutrinos arrived on Earth, many of them had changed by 'oscillating' from that flavour to another flavour, so they bypassed the detectors and went missing. The evidence for this happening was discovered and became ever more convincing between 1998 and 2001, and beyond, by the Japanese Kamiokande detector and the Canadian SNO.

Astronomers were proud that their meticulous work on the Sun had led to a new discovery about particle physics. The importance of this work was justly recognized by the award of the Nobel Prize in 2002 to Masatoshi Koshiba and Raymond Davis 'for pioneering contributions to astrophysics, in particular for the detection of cosmic neutrinos'. The director of the Sudbury Neutrino Observatory experiment, Arthur B. McDonald, was likewise awarded a share of the Nobel Prize in Physics in 2015 for the experiment's contribution to the discovery of neutrino oscillation.

The Crab Nebula
A supernova remnant

And in these Constellations then arise
New starres, and old doe vanish from our eyes:
As though heav'n suffered earthquakes, peace or war,
When new Towers rise, and old demolish't are.

John Donne, 'An anatomy of the World, the First Anniversarie', 1610

Imperial astrologers in China and Native Americans in the South-western United States recorded the appearance of a 'new star' in 1054. Nearly a thousand years later, Knut Lundmark and Edwin Hubble realized that these early observers had witnessed the birth of the Crab Nebula, a magnificent supernova remnant that has intrigued astronomers since it was first mapped in the eighteenth century.

Like the planet Uranus, the Crab Nebula was found as a result of a systematic whole-sky survey, conducted by an eighteenth-century British doctor, John Bevis, who had an observatory near London. In 1745 he compiled his observations into an atlas called *Urano-graphia Britannica*; the etched plates were costly to produce and the printer went bankrupt before printing it, so only a few proof copies survive. On the map of the constellation Taurus, near the star Zeta Tauri, Bevis drew a patch to represent a misty nebula that he had discovered.

The French astronomer Charles Messier used a copy of Bevis's atlas on his search for the predicted return of Halley's Comet in 1758. He found another comet, a new one, which passed through Taurus and drew his attention to the misty nebula. Comets and nebulae look much the same in a small telescope, so Messier, who was known as the 'Ferret of the Comets', decided to make a list of known nebulae to avoid confusion. The first item in Messier's catalogue was M1, Bevis's nebula, which became known as the 'Crab Nebula' after a bizarre sketch made in the 1840s by William Parsons, the Earl of Rosse, after viewing the nebula through his 'Six-foot' telescope at Birr Castle in Ireland. Modern pictures show the nebula as a generally oval shape of white light surrounded by a lacy network of filaments. The filaments are fragments of the body of an exploding star, and the white light comes from electrons spiralling around the star's magnetic field within the filaments.

The event in which the nebula first appeared was identified by the Swedish astronomer Knut Lundmark in 1931 as he listed the novae that had been recorded by Chinese astronomers and imperial historians. Chinese emperors maintained courts of astrologers who studied the sky in order to infer the future of affairs of state. Some of the signs that they recorded included 'guest stars' – temporary celestial phenomena, such as comets or novae. Generally, if the 'guest star' does not move for several days or months relative to the other stars it is probably a nova. Number 31 on Lundmark's list was a 'guest star' of July 1054. He noted that M1 was at the same position.

The clinching argument that connected the Crab Nebula with the guest star was American astronomer Edwin Hubble's measurement of the speed at which the M1 nebula is expanding, growing in size as its filaments rush continually outward from the centre of the explosion. Extrapolating backwards, Hubble found that in 1054 the filaments had been gathered together at the centre, and, in a series of popular essays, pointed out the correspondence with the Chinese record of the nova that had been noted by Lundmark. But Hubble's conclusion was overlooked until 1942 when it was revived by astronomers Jan Oort and Nicholas Mayall, and a Dutch Sinologist, Jan Duyvendak, who identified other historical records of the nova of 1054 from Korea, Japan and Baghdad.

In 1955 American astronomer William Miller proposed that a number of ancient rock paintings from Arizona and New Mexico depict the event. For example, on a roof ledge of a cave (now partially collapsed) a member of the Anasazi people living in Chaco Canyon, New Mexico, drew an image of the crescent Moon and a bright star, signing the picture with a handprint. Archaeological dating of the occupancy of the site is a rather long period of time (200 years), but it overlaps the date of the supernova, which was indeed seen in association with the crescent Moon. The evidence is circumstantial, and there are some discrepancies (the crescent is often the wrong way round), but it is a nice story and the star is reckoned by many to be an eyewitness representation of the supernova of 1054.

The Chinese astronomers had compared the brightness of the guest star of 1054 to other celestial objects like Venus. This made it possible to draw a light curve of the nova and show that it was in fact a supernova, a stellar explosion in which nearly the whole star disintegrates, leaving a black hole or a small stellar cinder, a neutron star. In 1968, while searching for the stellar remnant in the Crab Nebula with a radio telescope at Green Bank, West Virginia, radio astronomers David Staelin and Edward Reifenstein discovered a radio pulsar right in the middle of the nebula – a neutron star spinning on its axis thirty times per second. This was a brilliant confirmation of the bold idea put forward thirty years earlier by Fritz Zwicky that supernovae produce neutron stars.

The Crab pulsar flashes thirty times per second, and is a rotating neutron star with a diameter of about 29 kilometres. A strong magnetic field is embedded in the neutron star, and generates high-speed electrons that emit radio waves, so that the Crab Nebula is one of the brightest radio sources in the sky, having the designation Taurus A. Because the pulsar is losing energy, its rotation speed is gradually slowing. Occasionally the spin rate suddenly changes, speeding up but then recovering its long-term course. These so-called 'glitches' are an effect of the structure of the neutron star. The star has a crust that rotates at a slow rate compared with the interior, lagging behind. Occasionally, the crust cracks, the interior fastens onto the crust and the neutron star abruptly spins up.

The range of the scales of distance and time in the Crab is remarkable. The pulsar is 29 kilometres in diameter; the Crab Nebula 100 million, million kilometres. The pulsar spins round in one thirtieth of a second; the Crab supernova explosions occurred 30 billion seconds ago. The range of phenomena in the Crab is equally remarkable, to the extent that German-American astronomer Walter Baade suggested that astronomy was divided in two parts – the Crab Nebula, and everything else. It is a living textbook of astrophysics.

Planetary Nebulae
Looking into secret places

> On the evening of the 29th of August, 1864, I directed the
> telescope for the first time to a planetary nebula in Draco
> [NGC 6543]. The reader may now be able to picture to
> himself to some extent the feeling of excited suspense,
> mingled with a degree of awe, with which, after a few
> moments of hesitation, I put my eye to the spectroscope.
> Was I not about to look into a secret place of creation?

Sir William Huggins, *The New Astronomy: A Personal Retrospect*, 1897

One August evening in 1864, English astronomer William Huggins used his spectroscope to examine a nebula that he thought was a collection of densely packed stars. To his surprise, the spectrum showed 'a single bright line only!…The riddle of the nebulae was solved. The answer, which had come to us in the light itself, read: Not an aggregation of stars, but a luminous gas.'The nebula actually marked the quiet demise of a star much like the Sun, on its way to becoming a white dwarf.

Planetary nebulae are shells of illuminated gas that surround old stars. They were discovered by the English astronomer William Herschel, who in 1785 referred to the 'planetary', or 'disc-like', appearance of the object we now call NGC 7009, or the Saturn Nebula – it reminded him of the appearance of the planet Uranus. In 1790 he wrote that it was 'a most singular Phenomenon! A star… with a faint luminous atmosphere, of a circular form', adding that there could be 'no doubt of the evident connection between the atmosphere and the star.' He later discovered a similar nebula, NGC 6543, which also had a central star.

Herschel speculated that planetary nebulae were 'generating stars' (that is, stars in the process of being born) and that 'the further condensation of the already much condensed luminous matter may be complete in time.' He was articulating what became known as the Nebular Hypothesis of the origin of stars. He was actually completely on the wrong track: planetary nebulae are formations associated with the end of the life of stars, not their birth.

The gaseous nature of the nebulae was established by William Huggins. He became a pioneer in stellar spectroscopy after selling the family business at the age of thirty to pursue his interest in astronomy. In August 1864 he examined the planetary nebula in the constellation Draco (now identified as NGC 6543, or the Cat's Eye Nebula) with his spectroscope. He saw a single green emission line, which was reminiscent of the spectral lines that he had seen in gas discharge tubes in his laboratory. He had discovered that planetary nebulae are made of low-density gas. Later he identified other spectral lines in the nebula, some of which were generated by hydrogen.

In 1918 Lick Observatory astronomer Heber Curtis found that all planetary nebulae had central stars, which were visible if photographs could be taken at sufficient depths inside the nebulae. The stars are all white-hot, emitting copious amounts of ultraviolet light, which turned out to be key to answering the question of why the planetary nebulae shine. The mechanism was discovered by the Dutch astronomer Herman Zanstra in the 1920s, while he was a post-doctoral fellow. He showed that the amount of ultraviolet light emitted by the central star of a planetary nebula was identical to the amount emitted by the gaseous parts of the nebula: the energy from the star and the energy from the nebula matched. Every ultraviolet photon (light particle) emitted by the central star ionized one atom of hydrogen in the gas. When the ionized atom recombined, it emitted one photon of visible light.

Zanstra's theory accounted for the most abundant element in the nebulae, namely hydrogen, but other spectral lines remained unidentified. Lick Observatory astronomers William Campbell and James Keeler had identified some as being from helium soon after that element was discovered in the Sun in the late nineteenth century; Huggins had attributed other lines to a new element that he called 'nebulium'. But Henry Norris Russell, the director of the Princeton Observatory, remarked on the failure to reproduce the 'nebulium' lines in the spectrum of any other material that had been investigated, and concluded that they 'must be due not to atoms of unknown kinds but to atoms of known kinds shining under unfamiliar conditions. The suggestion is tempting that the nebular lines may be emitted only in gas of very low density.'

The American physicist and astronomer Ira Bowen confirmed Russell's guess in 1927 by discovering that in certain circumstances in space, common elements such as oxygen and nitrogen can emit spectral lines that they do not generate in the laboratory. Such spectral lines are described as 'forbidden'. On Earth, oxygen atoms never get a chance to emit these spectral emissions because collisions with other atoms interrupt the process, but in space the time between collisions is longer and the atoms have sufficient time to release their radiation. The gradations of colour in the nebula illustrate the decrease in temperature of the gases away from the centre.

The central stars of planetary nebulae are very hot, so they look rather faint because most of their energy is radiated as invisible ultraviolet. During the formation of planetary systems, the central stars are typically at the stage in their life cycles where they are transitioning from red giants to white dwarfs. The Ring Nebula (M57) is centred on a faint star that is on its way to becoming a white dwarf and has a surface temperature of 120,000 K.

The curious shapes and colours of the dust clouds in planetary nebulae inspire fanciful names. The Hubble Space Telescope has a beautiful gallery of images of planetary nebulae (plates XXIII and XXIV). Some planetary nebulae, like the Ring Nebula (M57), look circular and may be spherical. Another explanation for the shape is that we are viewing it down the axis of a three-dimensional barrel-like structure. Other planetary nebulae have complex shapes, often with some sort of symmetry like an hourglass (these are called 'bipolar nebulae'). Even some circular planetary nebulae like the

Owl Nebula have double features (like the 'owl's eyes') that suggest they are bipolar nebulae seen mostly end on. The Cat's Eye Nebula (NGC 6543) is more complicated. It is basically a bipolar structure lying within a series of spherical shells, all centred on a star turning from a red giant to a white dwarf. When it was a red giant, the star ejected some of its body into space at 1,500-year intervals, like a dog shaking dry its coat after a swim. The Butterfly Nebula (NGC 6302) is a bipolar planetary nebula, the two 'wings' of the butterfly shape being the two lobes of the bipolar shape. Its central star is hidden by an equatorial disc, which, unusually, contains significant quantities of icy dust. It is a mystery how this cold material has survived for so long so near a star that appears to have a temperature in excess of 250,000 K. It could be that discs like this play a part in making the bipolar shape of planetary nebulae, throttling the outflow from the red giant stage of their central stars.

The Origin of the Stars and the Planets

The solar nebula, proplyds and planetesimals

This world was once a fluid haze of light,
Till toward the centre set the starry tides,
And eddied into suns, that wheeling cast
The planets...

Alfred, Lord Tennyson, 'The Princess', 1847

The discovery of the origin of the Sun and its Solar System, and of other stars and planetary systems, is a story of inspired guesses. The theoretical investigations of an international who's-who of astronomers and astrophysicists dating back over 300 years were gradually proved by observations with spacecraft and detectors. The Sun and the Solar System formed from the collapse of part of a very dense, very cold cloud of interstellar dust and gas. The main part of the cloud condensed to become the Sun; orbiting lumps gathered surrounding gas and dust and grew into planets.

The explanation of the origin of planetary systems, first put forward by the Swedish scientist Emanuel Swedenborg in 1734 and the Prussian philosopher Immanuel Kant in 1755, is called the 'nebular hypothesis'. The nebular hypothesis was given support as a theory by the French mathematician and astronomer Pierre-Simon Laplace in 1796. He had proved that the Solar System was stable, with the orbits of planets oscillating by small amounts within their average values. The current shape of the Solar System, in which the planets all orbit in the same direction around a flat disc, reflects the way that it originally formed. In his book *Exposition du système du monde* (The System of the World), Laplace put forward the idea that the planets condensed out of a flat nebula that was whirling around the Sun.

Laplace added, in support of his theory, the observations by the English astronomer William Herschel in 1786 of nebulae that showed a single star embedded in the centre of a nebula. Herschel interpreted these as planetary systems seen in an early stage of development. In a striking image, Laplace compared them to the saplings in a forest of mature trees. In fact, the objects discovered by Herschel were not embryonic planetary systems at all – the first such system was discovered by Caltech astronomers Eric Becklin and Gerry Neugebauer in 1966. They used a newly developed infrared detector in a laborious scan, point by point, of a region of the Orion Nebula. During this scan they found a strong source of infrared radiation. Invisible to the unaided human eye, the 'BN Object' is the size of a planetary system. The infrared radiation comes from dust shrouding a new-born star inside. The dust traps the star's radiation and is heated to a temperature of about 700 K – at this temperature the dust radiates most strongly in the infrared spectrum.

The InfraRed Astronomy Satellite (IRAS) discovered further examples of proto-planetary systems in 1983, including discs of warmed dust grains orbiting the stars Vega, Zeta Leporis and Beta Pictoris. The disc of Beta Pictoris was photographed by Paul Kalas and Dave Jewitt in 1996, using a special camera attached to a small telescope on Hawaii – the clear skies at this high-altitude site concentrated the starlight into a small area. The starlight in this small area was blocked by a central obstruction. This made it possible to see the faint light from the dust.

The first direct images of planetary systems were made with the Hubble Space Telescope in 1992 by Robert O'Dell of Rice University and his colleagues. The images showed dust discs silhouetted against the luminous background of the Orion Nebula and other nebulae. O'Dell's wife named the objects 'proplyds' – a contraction of 'proto-planetary discs'. Their dust is concentrated in a disc rotating around a central star, just as Laplace visualized the solar nebula.

The way that the central mass of a proplyd contracts as a proto-star was first calculated by the Japanese astrophysicist Chushiro Hayashi in 1960. The star system does not immediately settle down but goes through paroxysms during which it ejects a stellar wind in every direction. It might also squirt jets of material from its poles. The material is ejected as a result of the rapid spin that the star has acquired as it contracts from the interstellar cloud from which it has formed. The slow rotation of the proplyd suddenly increases, just as, in a final flourish at the end of their dance routine, an ice skater will spin faster as they bring their arms closer to their body. The ice skater then brakes by grinding her skates into the ice; the star brakes by throwing off material. This quickly clears out a lot of the nebula in the neighbourhood of the newly formed star. But before it is cleared away, the nebula makes planets, which are massive and not so susceptible to the scouring force of ejected material from the new-born star.

The nebula surrounding the condensing star is mostly made up of hydrogen and helium. It contains dust grains that were made in old stars. These dust grains found their way into interstellar clouds, and thence into the nebulae orbiting the proto-stars – these are the dust grains discovered by astronomers in the BN Object, and imaged by the Hubble Space Telescope. Some of the material in the gas cloud assembles into molecules, which condense as ice on the surface of the dust grains. Still other material condenses into crystals. When the star switches on its nuclear reactions, it radiates energy, which melts the ice from the dust and blows away more gas in the inner, warmer parts of the solar nebula.

The dust that is left collides during its revolution round the star and sticks together, consolidating into larger lumps, centimetres to metres in size. According to the calculations proposed in 1969 by the Russian theorist Viktor Safronov, and in 1973 by Caltech theorists Peter Goldreich and William Ward, the disc then condenses further, with the lumps flowing past one another in streams that merge like the wake water behind a boat. This forms lumps that are kilometres in diameter. These lumps are called 'planetesimals'. Some of them survive in the Solar System as comets and asteroids. The gravity of the planetesimals is high enough to attract others – a large planetesimal is better able to do this. The more a planetesimal grows, the more its gravity increases, and the faster it grows – it is an accelerating process, called 'accretion'. The bodies formed in this way are called 'proto-planets'. Some of them survive as larger asteroids.

Our inner Solar System once contained about a hundred proto-planets formed by accretion. The accretion process stops when the proto-planets have emptied their immediate neighbour-hood of raw materials. Small fragments that are left over from all this are called chondrites – some fall to Earth from time to time as meteorites. Astronomers estimate their age by looking at the radioactive elements that are trapped in the meteorite material. This is the main way that astronomers establish the age of the planets, the Solar System and the Sun. It is a surprisingly accurately known number – 4.555 billion years old.

The larger gas planets in the outer Solar System formed by the same process, although material in the more distant part of the gas cloud is less affected by the heat of the star. Jupiter grew the fastest and the most, and its gravity disturbed the orbits of the inner proto-planets, some of which jaywalked across the main stream or revolution of the others and collided. These collisions caused some proto-planets to aggregate with the terrestrial planets – Mercury, Venus, Mars, and the Earth and Moon. Other collisions shattered proto-planets, creating the most common type of aster-oid. The fragments of shattered proto-planets rained down in the 'Heavy Bombardment' and made the large craters on the Moon and Mercury. One impact produced not one single planet, but a 'twin planet', the Earth and the Moon.

Interstellar Dust
Curtains of diamonds and graphite

> Until a person has thought out the stars and their
> interspaces, he has hardly learned that there are things
> much more terrible than monsters of shape, namely,
> monsters of magnitude without known shape. Such
> monsters are the voids and waste places of the sky.

Thomas Hardy, *Two on a Tower*, 1912

Imagine a cathedral, with a shaft of sunlight shining through the window. Specks of dust are floating in the sunlight. Then imagine the cathedral cleaned so scrupulously that there is only one speck of dust inside it. This represents the density of grains of dust that float in interstellar space, oxygen- and carbon-rich material expelled from supernovae and the interiors of red giant stars. There is not much dust in space – but there is a lot of space. The number of cathedral-sized volumes that stack one behind another in the line of sight to a star is very large, so the individual dust grains can accumulate to an opaque screen. Interstellar space is indeed filled with stardust.

In 1847, the Prussian astronomer Wilhelm Struve, who was working at the Tartu Observatory in what is now Estonia, first proposed that something inhabited the space between the stars. He had discovered that the number of visible stars per unit volume in the Galaxy decreases with distance from the Sun. He inferred that the light from distant stars was being absorbed by something in space. Dutch astronomer Jacobus Kapteyn discovered in 1909 that bluer stars moved across the sky more quickly than redder ones. Fast-moving stars are on average closer than slower ones, so Kapteyn concluded that the red stars were more distant, and not only dimmed but also reddened by interstellar absorption, much as dust in the lower atmosphere of the Earth reddens the setting Sun. Similar work was carried out in 1930 by Robert Trumpler, then at the University of California's Lick Observatory, on clusters of stars – he found that the clusters with smaller diameters are more distant than the larger ones, and fainter than their distances alone would account for because of interstellar absorption by dust.

In the first two decades of the twentieth century, the American astronomer Edward Emerson Barnard carried out a systematic programme to photograph our Galaxy. In his atlas of the Milky Way he identified distinct dark 'holes' in the star clouds. Astronomers since William Herschel had known of the existence of these holes, and for a long time thought they were true voids in the distribution of stars. But Barnard discovered that the holes were 'obscuring bodies nearer to us than the distant stars' – dark clouds of unusually dense interstellar dust.

These dust clouds concentrate towards the plane of the Galaxy, which is why the Milky Way appears to be cleft along its central line when we view it edge-on from Earth. One of the most prominent clouds lies in the Southern Cross and is called the Coalsack. In the culture of some Australian aboriginal peoples, the Coalsack represents the head of an emu defined by the straggling form of the Milky Way between Crux and Scorpio, a unique 'constellation' made of dark dust clouds rather than stars.

Some of the dark clouds are very small. IC 2944 is a star-forming region in Centaurus. Silhouetted on the nebula are dense, opaque clouds of interstellar dust, first spotted by South African astronomer A. D. Thackeray from the Radcliffe Observatory, Pretoria, in 1950 and termed 'Thackeray's globules'. They may collapse to form proplyds or they may be shredded by intense ultraviolet radiation from the young, hot stars and disperse, rather than forming new stars themselves.

Dust grains that lie near to bright stars can reflect starlight and form a 'reflection nebula'. There is a prominent example in the

Pleiades star cluster, whose stars illuminate a dark dust cloud that they encountered as they coasted through space. The nature of this nebula, the first reflection nebula found, was discovered by Lowell Observatory director Vesto Melvin Slipher in 1913, who observed the spectrum of the nebula and found that it was identical to the spectra of the brighter Pleiades stars. The Solar System originally formed from interstellar gas and dust. High temperatures destroyed most interstellar dust particles in the solar nebula, but some meteorites (known as carbonaceous chondrites) contain small particles whose composition is not the same as the rest of the meteorite, and which are thought to be interstellar grains. These were convincingly identified in 1987 by University of Chicago physicists Ed Anders, John Wacker and Tang Ming, and Washington University physicist Ernst Zinner, who isolated interstellar diamond and silicon carbide in meteorites by dissolving the rest of the meteorite in acid, a method referred to as 'burning down the haystack to find the needle'.

The Ulysses and Galileo spacecraft also detected interstellar grains in the Solar System, using a microphone to detect impacts by interplanetary dust particles. Venturing beyond Jupiter, Ulysses and Galileo encountered a higher than expected number of impacts coming from a particular direction in space and hitting the spacecraft at the same speed. They were from a stationary cloud of interstellar dust particles through which the Solar System is moving at 20 kilometres per second. Before this, it had been thought that the solar wind would stop interstellar dust grains from entering the Solar System, but we now know that the larger grains manage to break through.

Some of these interstellar particles have been brought back to Earth by the Stardust spacecraft, which between 2000 and 2004 deployed a sticky gel in interplanetary space and at Comet 81P/Wild (also known as Wild 2, pronounced 'Vilt Two') and returned to Utah in January 2006, so that the material that had been collected on the gel could be analysed. Most particles were from the comet, but some have proved to be interplanetary and interstellar grains. The types of interstellar dust that have been discovered include tiny diamonds and larger graphite grains that are made in supernovae, as well as silicon carbide, aluminium oxide, spinel and titanium oxide grains made in the atmospheres of red giant stars before they turn into planetary nebulae.

Interstellar grains are little factories where molecules are built in space. They have sticky surfaces, with atomic-scale electrical charges that provide attractive 'hooks' that latch onto passing atoms in the interstellar gas. When two such atoms – for example two hydrogen atoms (H + H) – are brought together side by side on the surface of a grain, they may loosen their hold on the grain and clutch each other to make a molecule – for example molecular hydrogen (H_2) – giving up excess energy to the grain. The hydrogen molecule will then drift off the dust grain and, eventually, become part of a giant molecular cloud, and participate in the formation of stars and planetary systems. Like termites in the vast areas of the savannah, interstellar grains, small but numerous as they are, play an important part in the ecology of space.

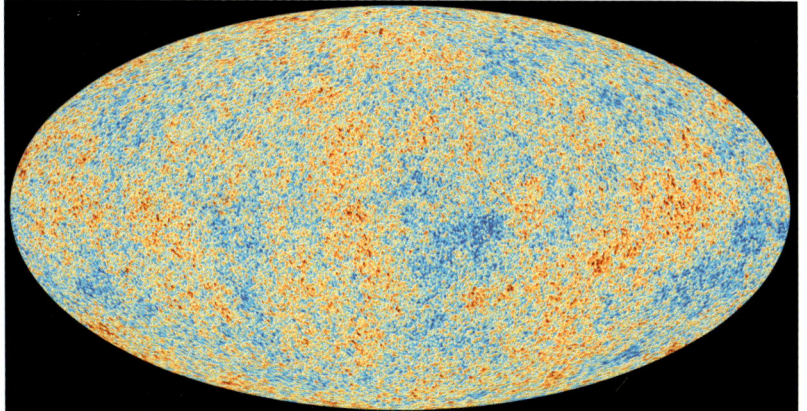

XVI The Cosmic Microwave Background. The Planck satellite made this image of the remnant fireball of the Big Bang.

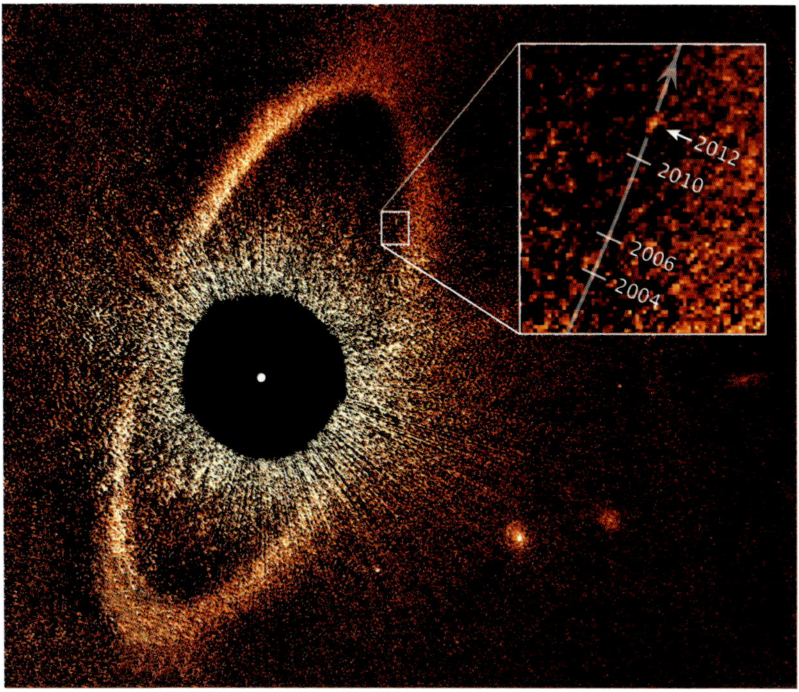

XVII The star Fomalhaut is blocked out in this image by the Hubble Space Telescope, so that we can see the dusty disc that surrounds it and the planet that orbits it, shown at four positions that span eight years of its 2,000-year orbital period.

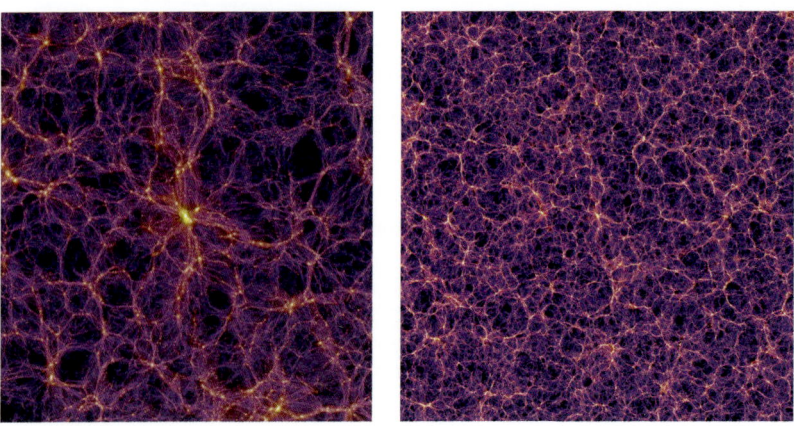

xviii Cassiopeia A. This remnant of a supernova that exploded 300 years ago emits light (Hubble Space Telescope image, orange), infrared (Spitzer Space Telescope, red) and X-rays (Chandra X-ray Observatory, blue and green).

xix The Millennium Simulation. Matter, dark matter and dark energy are mixed in a computer calculation to show how they evolve and generate galaxies, linked together in a 'cosmic web'.

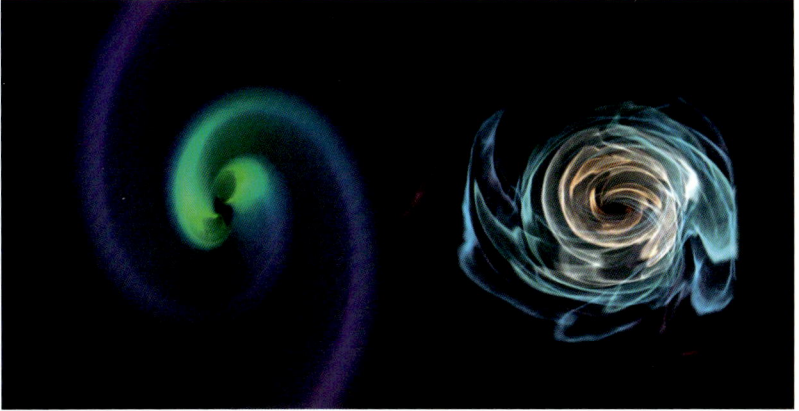

xx Saturn's rings. The orbiting particles that constitute Saturn's rings are colour-coded by size (green less than 1 centimetre, purple from 5 centimetres up to several metres across).

xxi Gravitational waves. Two orbiting neutron stars coalesced in 2017 and emitted a spiralling pattern of gravitational waves, detected by the LIGO interferometer (left). The matter from the neutron stars is shown right, ejected in a more chaotic pattern.

XXII The Orion Nebula. New-born stars embedded within a giant molecular cloud have blown a cavity whose inner surface they illuminate. Atoms of gas glow with spectacular colours.

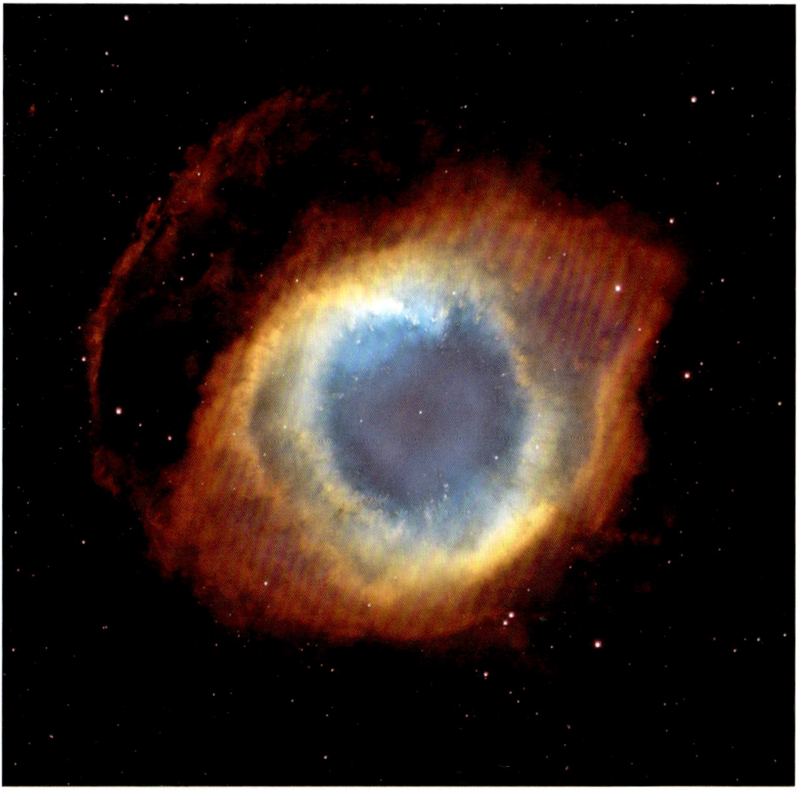

ABOVE

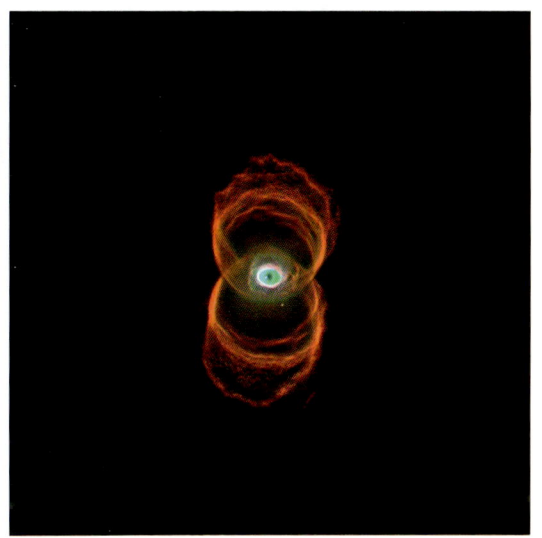

XXIII The Helix Nebula. This dramatic planetary nebula is centred on the star that produced it. The star is making the transition from red giant to white dwarf.

LEFT

XXIV MyCn18. This young planetary nebula is shaped like an hourglass with an intricate pattern of 'etchings' in its walls.

ABOVE

xxv The Antennae. Two galaxies have collided, with their gas surging together and forming a sparkling starburst of bright new stars.

RIGHT

xxvi The Whirlpool Galaxy, M51. The first galaxy in which the characteristic spiral shape was recognized has a small companion at the end of one spiral arm.

xxvii Centaurus A. Lobes and jets emanate from the galaxy's central black hole. The image is a composite of images obtained with infrared (orange), X-rays (blue) and light (true colour).

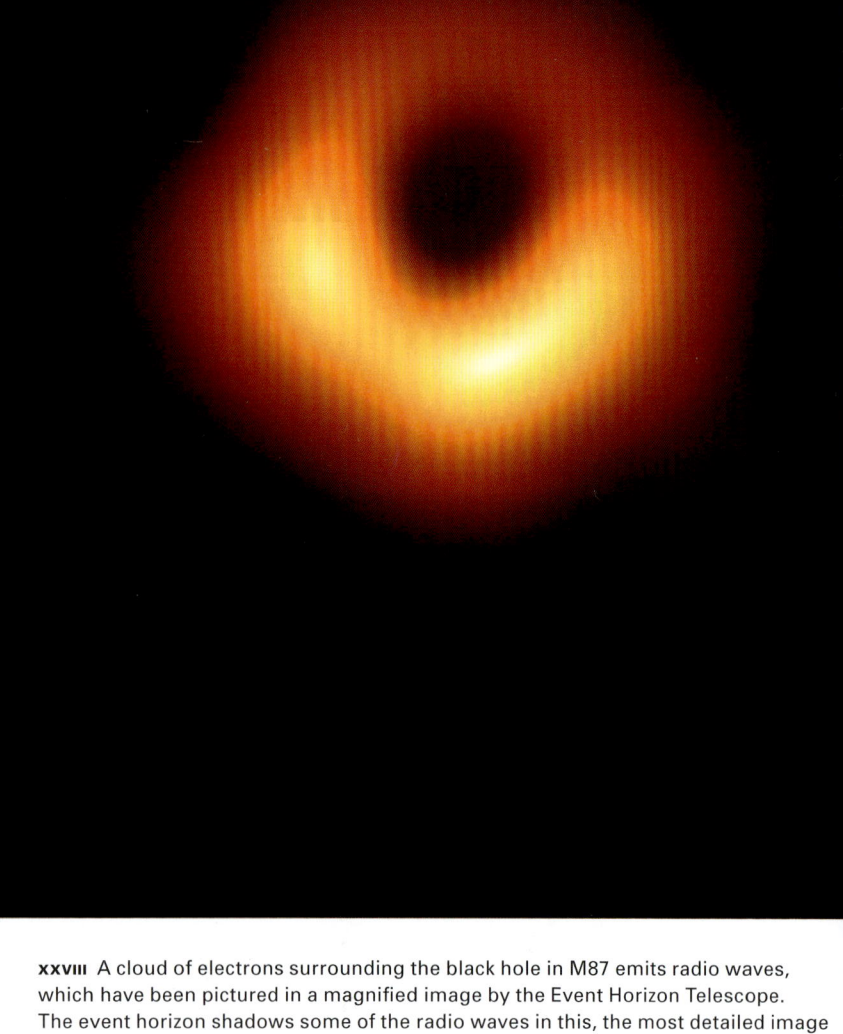

xxviii A cloud of electrons surrounding the black hole in M87 emits radio waves, which have been pictured in a magnified image by the Event Horizon Telescope. The event horizon shadows some of the radio waves in this, the most detailed image yet taken of a black hole.

THE UNIVERSE

AND ITS

GALAXIES

Hydrogen
The most abundant element in the Universe

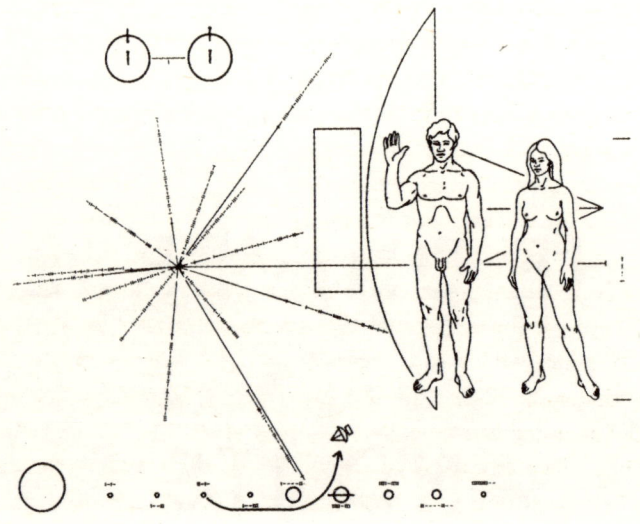

Credit: NASA Ames

Carl Sagan, Frank Drake and Linda Salzman Sagan,
Plaque from the Pioneer 10 spacecraft, 1971

Hydrogen is the most abundant element in the Universe. It was made in the Big Bang, and condensed to form large-scale gaseous structures, which in turn produced galaxies and stars. Our own Galaxy contains hydrogen both in stars and in interstellar space. Under the shadow of Nazi occupation, a group of Dutch astronomers worked in secret to identify the radio signature of this interstellar hydrogen. After the war, their discoveries would make it possible to map the entire Galaxy.

There are three types of interstellar hydrogen in space. One type (called 'H II') is ionized hydrogen, which is present in areas where stars are forming and becomes visible to the naked eye when it is excited by ultraviolet light emitted by hot stars. By locating clouds of ionized hydrogen, William W. Morgan, Stewart Sharpless and Donald Osterbrock were able to plot the location of nebulae in a 1951 map of the Galaxy, which was the first to show its structure and scale.

Outside the nebulae, away from the stars that are sources of ultraviolet light, are vast amounts of a cool, invisible form of non-ionized hydrogen gas called 'H I', or neutral hydrogen. The existence of neutral interstellar hydrogen was confirmed by radio astronomers in the 1950s, following a remarkable prediction made by a group of astronomers who met in secret during the Nazi occupation of the Netherlands. Their work was published only after the Second World War ended in 1945.

When Germany invaded the Netherlands, Hendrik van de Hulst was an astronomy student studying under the astrophysicist M. Minnaert in Utrecht. Minnaert was sent to a detention camp for the duration of the war after protesting the treatment of his Jewish academic colleagues and Van de Hulst fled to Leiden, where he studied under Jan Oort. Oort encouraged his astronomical students to concentrate on theoretical studies while the war was under way, since observing with telescopes at night while the curfew was in effect caused suspicion on the part of the occupying forces, and was therefore highly dangerous. Through Bart Bok, a Dutch astronomer working in the USA, copies of Grote Reber's papers on radio astronomy had been smuggled into the country, and in the spring of 1944 Oort decided to hold a colloquium on Reber's findings.

Although the occupying forces had banned most public gatherings, to counter conspiracy and resistance, Oort arranged a meeting of the Astronomenclub (Dutch Astronomy Club), where Van de Hulst presented calculations to support his theory that hydrogen emits radio waves that have a distinct signature. He chose to study hydrogen because it was the most abundant element in the Universe.

Hydrogen atoms consist of an electron that is in orbit around a proton. Both the electron and the proton have a spin, and the axes of the spins can be parallel or antiparallel. Van de Hulst found that if the electron flips spontaneously from a parallel to an antiparallel spin, the hydrogen atom emits a pulse of radio waves that have a wavelength of 21 centimetres. Although these electron 'flips' happen only once every 11 million years in an individual hydrogen atom, the number of hydrogen atoms in space is so large that a hydrogen cloud produces measurable amounts of 21-centimetre radiation every second.

After the war had finished, Oort and his colleague Lex Muller attempted to prove Van de Hulst's theory by isolating the 21-centimetre radiation with a radio receiver, but suffered a setback when their equipment in Kootwijk was destroyed in a fire. The spectral line from the hydrogen radio emission was finally discovered in 1951 by American radio astronomers Harold Ewen and Edward Mills Purcell (later a Nobel laureate for his work on the fundamental physics of the hydrogen atom), using a radio telescope built at weekends by Purcell with a grant of only $500 (equivalent in purchasing power to about $5,000 today). This discovery was confirmed by Oort and Muller after they repaired their receiver, and subsequently reconfirmed by Australian radio astronomer Frank Kerr in Sydney. All the discoveries were published together as a set of three papers in the magazine *Nature*.

The first maps of the Galaxy, drawn according to Oort's prescription, were made by Van de Hulst, Muller and Oort in 1952. In 1958, Oort, Kerr and Dutch-American astronomer Gart Westerhout mapped virtually the entire Galaxy by combining observations of hydrogen radio emissions from astronomers in Australia and the Netherlands. They correlated the neutral hydrogen with the hot stars and nebulae mapped by Morgan, Sharpless and Osterbrock and confirmed that our Galaxy has spiral arms. Its structure is similar to the spiral galaxy M51 (plate XXVI), sketched in 1845 by William Parsons, the Third Earl of Rosse, as he viewed it with his 6-foot telescope, then the largest in the world and known as 'the Leviathan of Parsonstown'.

The third form of hydrogen found in interstellar space is in the form of hydrogen molecules (H_2), two hydrogen atoms joined together. It was considered likely to be common as far back as the 1930s, long before it was actually discovered. The molecule was observed first with space telescopes since its prominent signatures are masked from ground-based telescopes by molecules in the Earth's atmosphere. It was first seen in 1970, using instruments borne aloft on rockets. In 1972, the Copernicus satellite detected interstellar molecular hydrogen. Further advances in its study were made by the Far Ultraviolet Spectroscopic Explorer (FUSE, active between 1999 and 2007) and the Hubble Space Telescope. Molecular hydrogen is the most abundant molecule in interstellar space, and accumulates in Giant Molecular Clouds, the places within which new stars and planetary systems are born.

Hydrogen's 21-centimetre radio radiation is a universal phenomenon that would presumably be known to any scientific civilization on other planets in the Galaxy. It has thus featured in attempts to communicate with extraterrestrial beings. It has several times been chosen as the radio frequency with which to set up a radio link to such beings – one has to choose one wavelength out of the almost infinite number of possibilities in the radio spectrum, and the 21-centimetre wavelength has been reckoned as the one with the greatest cosmic significance. Additionally, this wavelength was used in the attempt to communicate with extraterrestrial beings through what amounted to sending a letter through the post, using the Pioneer 10 and 11 spacecraft as mail vans. Launched in 1972 and 1973 respectively, they are currently sailing out of the Solar System into interstellar space, carrying plaques with pictorial messages. The intention is that any extraterrestrial beings who might find a spacecraft would be able to get an indication of who made and launched it. The plaques (the design is reproduced at the head of this chapter) depict the location of the Solar System in the Galaxy, relative to a number of pulsars, and show that the spacecraft originated on the third planet from the Sun, having flown past the fourth, fifth and sixth planets. The plaques also depict two people, a man and a woman, the man greeting the finders and the woman passively looking on (the plaque was conceived at a time when feminism had not penetrated very deeply into NASA). The scale of the humans is indicated by the spacecraft and by a diagram of an atom of hydrogen (in the upper left corner of the plaque) emitting 21-centimetre radio waves as it flips its spin axis.

Galaxies
Ellipticals, spirals, mergers

I see beyond this island universe,
Beyond our Sun, and all those other suns
That throng the Milky Way, far, far beyond,
A thousand little wisps, faint nebulae,

...

Faint as the mist by one bright dewdrop breathed
At dawn, and yet a universe like our own;
Each wisp a universe, a vast galaxy
Wide as our night of stars.

Alfred Noyes, 'William Herschel conducts', 1922

Galaxies are in constant motion, speeding outwards as the Universe expands, occasionally colliding with each other and changing shape. Thousands of distant galaxies – representing only a fraction of the total in the Universe – were captured in an astonishing set of images called the Hubble Deep Fields, which reach to the frontier of the visible Universe.

Elliptical galaxies are smooth and featureless. In three dimensions they are triaxial ellipsoids – aspherical balls with unequal sizes along the three planes of symmetry. Spiral galaxies are flat discs with spiral arms. Lenticular galaxies are a transitional type between spirals and ellipticals. Irregular galaxies are small, without a clear shape, or they are larger and appear to be two galaxies passing close to each other and disrupting any regular shape (plate xxv). In other interactions, a small galaxy might be absorbed by another.

Mergers like this are going on now in our own Galaxy. The Sagittarius Dwarf Elliptical Galaxy was discovered in 1994 by Cambridge astronomers Rodrigo Ibata, Mike Irwin and Gerry

Gilmore, who noted an excess of faint stars grouped just above the plane of our Galaxy at a distance of about 70,000 light years. The Milky Way has disrupted this galaxy into a stream of stars that loop in orbit over the pole of our Galaxy and are merging with it. Our spiral Galaxy has grown by several such mergers in the past, and if it merges in the future with another spiral, it seems likely that both will lose all their spiral features and make an elliptical galaxy – in fact, astronomers James Binney and Scott Tremaine discovered in 1987 that the Milky Way galaxy will collide with and merge with the Andromeda spiral galaxy, M31, in about 2–5 billion years. Our Sun will likely be a red giant at that time. Its fate after the merger is unclear: it may well be flung into very lonely intergalactic space.

Edwin Hubble did not only classify galaxies, but also discovered their distances from Earth and determined how they were moving. With the 100-inch Mount Wilson telescope, he observed Cepheid variable stars in spiral galaxies, which told him their distances, far outside the Milky Way. He used measurements by Vesto Melvin Slipher of the speed of forty-six galaxies and in 1929 he discovered they were, in general, receding – moving away from our Galaxy. The galaxies' speeds were proportional to their distances. There was some scatter (the Andromeda Galaxy, for example, is approaching us, which is why it will merge with our Galaxy soon), but, according to Hubble's determination, the trend line showed that a galaxy at a distance of 3 million parsecs was receding at 500 kilometres per second. This is a factor of 8 too high, according to modern calibrations of the measurement of the distance of galaxies by Cepheids, but the principle remains: the Universe of galaxies is expanding, and the trend line is called Hubble's Law. It was the first indication that the Universe exploded in a Big Bang.

The most distant galaxies studied are radio galaxies and the galaxies in the Hubble Deep Fields and similar surveys. In 1995 the Hubble Space Telescope stared at an almost empty area of the Northern Hemisphere sky for ten days to take the deepest picture of the sky obtained up to that time. It did the same thing in 1998 with a similar area in the southern sky, to confirm whether the Universe in that direction was the same as in the north – it was. It capped its own efforts several times with additional deep fields made with cameras that were more sensitive, and sensitive to different radiations. These, the deepest (most sensitive) astronomical pictures ever made, contain the images of several thousand galaxies. Some were formed less than a billion years after the Big Bang.

The Hubble Deep Fields showed that, in general, galaxies as young as a billion years – the most distant galaxies which the photographs had discovered – are smaller than galaxies are today, 14 billion years after the Big Bang. This is because most galaxies around today have merged with others at some time in their lives – galaxies are now fewer but bigger. The galaxies in the Hubble Ultra Deep Field are also less symmetrical. They are younger and have not settled down to a steady state; they are seen in gawky adolescence rather than mature equanimity.

Not much further than the distance of the faintest galaxies in the Hubble Deep Fields, there are none visible. This time is called the Dark Ages of the Universe. Astronomers believe that galaxies exist beyond this point, but are cloaked in dust that hides the light of the stars inside. The dust is warmed by the stars and emits infrared and millimetre radiation, and so the galaxies show up only as faint sources of infrared or millimetre radiation. Several such extremely faint sources have been discovered by the SCUBA cameras on the James Clerk Maxwell Telescope on Mauna Kea, Hawaii. They are so faint that it is extremely difficult to discover much about their properties, even with the Hubble Space Telescope. But the prospects are favourable. Just as the Hubble Space Telescope was targeted at investigating optically visible galaxies, the James Webb Space Telescope (scheduled to be launched in 2021) is aimed at elucidating the nature of their predecessors.

Magellanic Clouds
Our neighbour galaxies

> [Our ship was at 37 degrees south latitude when] we sawe a
> marueylous order of starres, so that in the parte of heauen
> contrary to owre northe pole, to know in what place and
> degree the south pole was, we tooke the day with the soonne,
> and obserued the nyght with the Astrolabie, and sawe
> manifestly twoo clowdes of reasonable bygnesse mouynge
> abowt the place of the pole continually now rysynge and
> nowe faulynge, so keepynge theyr continuall course in
> circular mouying, with a starre euer in the myddest which is
> turned abowt with them abowte xi degrees from the pole.

Andrea Corsali, *Letter of 1516 to Giuliano de' Medici, Duke of Nemours*,
translated by Richard Eden in 1555

The outline of the Milky Way as seen from the Southern Hemisphere
is very irregular. Two pieces seem to have broken off; they are known as
the Magellanic Clouds, and appear in the mythology of many peoples
in Africa, Australia and South America. We now know that the 'clouds'
are actually two separate galaxies that are in orbit around our own.

The first known mention of the Magellanic Clouds is in the *Book
of Fixed Stars* (964 CE) by Persian astronomer Al-Sufi. He called
the Large Magellanic Cloud Al Bakr, the White Ox, and said that
while Al Bakr was invisible from northern Arab countries, because
it was so close to the south celestial pole, it could be seen from the
strait of Bab-el-Mandeb, which is the southern outlet of the Red
Sea into the Indian Ocean.

The Magellanic Clouds were first seen by Europeans during the
early voyages of discovery to the southern seas in the fifteenth and
sixteenth centuries. They became known as 'Cape Clouds', referring

to the Cape of Good Hope, from which they could be spotted. They were drawn with the Southern Cross on a star chart in 1516 by an Italian navigator and spy, Andrea Corsali, who travelled as a double agent for the Medici family, seeking out commercial possibilities for them on a secret Portuguese voyage to India.

Later the clouds became associated with Ferdinand Magellan (Fernão de Magalhães), the intrepid Portuguese captain who led the first voyage around the world between 1519 and 1522. They were noticed by members of his crew and their existence recorded in accounts of the voyage; unfortunately, Magellan himself had no opportunity to participate in the story of their discovery since he had been killed in the Philippines during the final months of the voyage. Antonio Pigafetta, an Italian navigator who sailed with Magellan, wrote after they had passed through what are now called the Straits of Magellan: 'The Antarctic pole is not so covered with stars as the Arctic, for there are to be seen many small stars congregated together, which are like two clouds a little separated from one another and quite dim, in the midst of which there are one or two stars...' Because they are near the south celestial pole the Clouds were useful navigation aids, much as the Great Bear is to sailors of the Northern Hemisphere. In the seventeenth century the Clouds were often called by their Latin names of Nubecula Major and Nubecula Minor (the Large and Small Magellanic Clouds, LMC and SMC, respectively).

The two Clouds were examined by John Herschel on his astronomical expedition to the Cape in 1834–38. He claimed to have seen a connection, or star drift, lying between them, and he certainly discovered and catalogued 244 star clusters, double stars and the like in the SMC and 919 in the LMC. The first suggestion that they were separate galaxies lying outside our own Galaxy was made (as one suggestion among three) by Cleveland Abbe, the US astronomer who, unable to find a job in astronomy, turned meteorologist, since this profession would be the greatest assistance to other astronomers. This became ever clearer through the studies set in train by the Boyden Observatory established in the Southern

Hemisphere by the Harvard College Observatory, first at Arequipa, Peru in 1889 and then moved to Bloemfontein, South Africa, in 1927. The pioneering female astronomer Henrietta Leavitt used data from these observatories to study variable stars in the Clouds and discovered the Period Luminosity relationship.

These galaxies are some of our nearest neighbours – the largest two of about two dozen satellite galaxies to our Galaxy. The fact that they are separate from our Galaxy, so that they can be seen as a whole while their contents can be seen in detail, makes them very important galaxies to study. For example, the properties of the individual star that exploded as Supernova 1987A in the LMC had been known before this happened, including its distance – the first time this had been possible. But the Clouds' positions at high southern latitudes mean that they cannot be studied very effectively from anywhere in the Northern Hemisphere. These considerations have meant that southern astronomers have concentrated on studies of the Magellanic Clouds. Using the Parkes radio telescope in New South Wales, Australian astronomers P. Wannier, G. T. Wrixon and Don Mathewson discovered the Magellanic Stream of hydrogen gas that links the two galaxies and our Galaxy. It was drawn out of the three galaxies by their interaction as they orbit around each other.

The nature of the orbit is being unravelled by Gaia, a space satellite from the European Space Agency that is measuring the nature and the motions of the stars of our Galaxy and beyond. It uncovered a complicated situation, because there are groups of stars in the galaxies that are being pulled off each galaxy by the ongoing interaction. There may even have been a third Magellanic Cloud that collided with the others in the past, so the Magellanic Clouds, currently a duo, might once have been a threesome in a relationship that has literally broken up. Another issue is whether the Magellanic Clouds are actually in repeated orbit around our Galaxy, or whether they are on their first and only approach, swingers just casually passing through.

Quasars
Active galaxies

[3C 273] had been occulted by the Moon several times and
Cyril Hazard et al. had obtained very accurate positions
in Australia...The identification was a 13th mag star and
a very faint jet-like feature. Convinced that the bright star
could not be the radio source, I obtained its spectrum in
December 1962. The spectra showed a number of broad
emission lines. Several weeks later...I realized that four of
the six emission lines showed a regular pattern of spacing
and intensity. Soon I realized that it was the Balmer
spectrum of hydrogen, redshifted by 16 percent. The same
day Greenstein and I found that 3C 48 had a redshift of
37%. It was a totally unexpected development. How could
a star exhibit a big redshift? It could be a cosmological
redshift like that of galaxies, but that would make its
luminosity a hundred times larger than the typical galaxy...

Maarten Schmidt, *Autobiography*, 2008

A quasar is an 'active galaxy': a galaxy that is exceptionally bright
because it emits most of its energy from its nucleus. This energy
is not starlight, but light and radio waves pouring from a massive
central black hole. Quasars were discovered in the 1950s when
astronomers tracking these strong radio emissions noticed a strange
bright star that proved not to be a star at all.

Quasars are exceptionally bright galaxies, emitting light and
radio waves from a central black hole of supermassive proportions
– billions of times the mass of the Sun. The black hole is feeding
on gas that infalls at speeds of many thousands of kilometres per
second. So much gas falls in towards the black hole and so much

energy is released that, on the interaction of the energy with the gas, some gas is turned around and ejected. As a result, many quasars show jets shooting out at high speeds, in two opposite directions.

Quasars are the brightest of so-called 'active galaxies'. The first to be discovered were spiral galaxies identified by American astronomer Carl Seyfert in 1943. He noticed their unusually bright nuclei and that they had strong emission lines in their spectra, coming from gas that sometimes moved very quickly. Only later did it become clear that the gas was in orbit around something very massive. At the time, the galaxies were enough of an unexplained curiosity to be given a new designation, 'Seyfert galaxies'.

In the next development, radio astronomers discovered that some galaxies emit radio waves – the first recorded radio galaxy was for some years an unremarked bump on Grote Reber's 1939 radio map showing the radio-wave Milky Way as it passes through the constellation Cygnus. If anyone thought about the bump at all, they regarded it as a feature of our Milky Way galaxy. Then, in 1946, British physicist John Hey and his colleagues used surplus radar equipment to study this source, which they named Cygnus A, as the strongest in the con-stellation. It was very small and some astronomers thought that it was a new kind of 'radio star'. Others, including Thomas Gold and Fred Hoyle, argued that it was not a star but an object outside the Milky Way galaxy. 'Why...does not one [person] find any identifiable visual object where those very near radio stars are supposed to be?' asked Gold. He pointed out in 1951 that the fifty radio sources known then did not concentrate towards the Milky Way like the stars in our galaxy, but were uniformly scattered over the sky, like other galaxies are.

Later in 1951, Cambridge radio astronomer Francis Graham Smith measured the radio position of Cygnus A accurately enough to make it worthwhile to try to find out what was at the same place in the visible sky. Smith airmailed the position to astronomer Walter Baade at the California Institute of Technology in Pasadena. In April 1952 Baade took two photographs with the 200-inch Mount Palomar telescope. 'There were galaxies all over the plate, more than two hundred of them, and the brightest was at the centre. It showed signs of tidal distortion, gravitational pull between the two nuclei – I had never seen anything like it before. It was so much on my mind that while I was driving home for supper, I had to stop the car and think.' Baade concluded that Cygnus A was two galaxies in collision.

Discussing his discovery with a sceptical colleague, Ralph Minkowski, Baade bet him that the spectrum of Cygnus A would have spectral emission lines from highly energetic gas, the stake for the bet being a bottle of whiskey. Minkowski soon took the spectrum with the Palomar telescope and conceded the bet, although he need not have done, since the emission lines come not from a collision between galaxies but from the massive black hole in Cygnus A.

Radio-astronomy technology improved and Cambridge radio astronomer Martin Ryle invented the powerful kind of radio interferometer for which he later received the Nobel Prize. Radio engineers in Cambridge and in Sydney, Australia, competed to build successively more and more sensitive and accurate instruments. Radio surveys of the northern and southern sky with these telescopes from 1953 onwards discovered vast numbers (several thousands) of galaxies that emit radio waves. One comprehensive and accurate catalogue published in 1959 and revised in 1962 received the designation 3C (the third such catalogue made in Cambridge) and 3CR.

A strong source in the catalogue had the number 3C 273. It lay in the band of the zodiac and from time to time the Moon passed in front of it and occulted it. British radio astronomer, Cyril Hazard, used the newly built Parkes radio telescope in Australia to watch 3C 273 during a series of occultations of the radio source by the Moon in 1962. He was able to pin down its position by plotting the edge of the Moon at the moment that 3C 273 disappeared – he noticed that it disappeared in two steps and must be double. Tom Matthews found that the double radio source coincided with what looked like a 13th magnitude star and a small wisp or jet attached to its image. Suspecting that the wisp was a galaxy and the source of the radio waves but calculating that it was too faint to study, Caltech astronomer Maarten Schmidt took a spectrum of the 'star' in order to eliminate it. It was very bright for the 200-inch telescope and his first attempt was overexposed. Within days, however, Schmidt had discovered that 3C 273 was not an ordinary star – it had spectral lines in emission, indicating hot gas was present, but he could not identify the emission lines with anything he had seen before, though he tried several different sorts of explanations. Nor could they be identified even by world experts to whom Schmidt showed the spectrum.

Collaborating with Cyril Hazard in writing up all the work on 3C 273, Schmidt tried to systematise the wavelengths of the lines in a diagram and suddenly noticed that four of them formed a progression that reminded him of the spectrum of hydrogen – but with the wavelengths red-shifted by a huge factor. When he applied the same factor to the other spectral lines, their identification made immediate sense. Schmidt had discovered that the bright, star-like radio source 3C 273 was a galaxy at a huge distance – the technical names Quasi-Stellar Object or Quasi-Stellar Radio Source came to be abbreviated as QSO or quasar. The tiny size of 3C 273 was apparent in 1961 when Harlan Smith and Dorrit Hoffleit looked back through the archive of sky photographs at the Harvard College Observatory and discovered that 3C 273 varied by large amounts on a timescale of years. This meant that it could only be at most light years in dimension, contrasted with the size of a normal galaxy, which is many tens of thousands of light years in size. Incredibly bright, at incredible distances, incredibly small – this was the paradox of the quasars.

Because quasars are so luminous, they control the birth of stars in their host galaxy: the birth process turns off when the quasar increases its brightness. Quasars even affect the entire Universe by ionizing intergalactic hydrogen gas. The most distant quasars discovered were born only a few hundred thousand years after the Big Bang, and astronomers still struggle to explain how material amounting to billions of times the mass of the Sun can gather together to form a black hole in so short a time.

Supermassive Black Holes
Monsters at the centres of galaxies

Therefore, some are in darkness;
Some are in the light, and these
You may see, but all those others
In the darkness no one sees.

Bertolt Brecht and Kurt Weill, *Threepenny Opera*, 1928, translated
by Christopher Isherwood and quoted by Engelbert Schucking in
'Kinematics of relativistic ejection', 1976

Nature makes black holes in two scenarios: in the aftermath of a supernova explosion, or in the nucleus of an active galaxy. 'Active galactic nuclei' (AGNs) is the generic name for quasars, radio galaxies, Seyfert galaxies and the like. They are all supermassive black holes: black holes that are much more massive than stars, and lurk in the centres of galaxies (plate xxvii).

Although AGNs are astoundingly bright, they are tiny, perhaps only light hours in diameter. Gas in AGNs moves very quickly (as fast as 100,000 km/s), so there must be some kind of massive, compact structure around which the gas orbits. In fact, in 1964 Russian astronomers Yakov Borisovich Zeldovich and Igor Novikov calculated that if AGNs were not massive enough to generate a strong gravitational force, their intense radiation would blow them apart. The huge mass of AGNs was confirmed in 1994 by a team led by Holland Ford, who used the Hubble Space Telescope to discover that the nucleus of the AGN known as M87 was 3 billion times the mass of the Sun. Some AGNs, like 3C 273, have jets that are long, narrow and straight, and stay pointed in the same direction for millions of years. One explanation is that when the jet shoots out from a rotating body, along its axis, it acts as a stable gyroscope that maintains its own direction.

All these clues told astronomers Edwin Salpeter in the USA and Yakov Zeldovich in the USSR what the compact structure in quasars could be – a rotating, supermassive black hole. The key breakthrough came in 1969 when English astrophysicist Donald Lynden-Bell put clothes on the black hole theory. He argued that a quasar's energy came from frictional heating of a gaseous orbiting disc of material: the inner bits of the disc orbit faster than the outer bits, and they scrape together. Astronomers had already seen evidence for the disc – the rapidly moving material – and found that the spectrum of the disc was just as Lynden-Bell predicted. So all AGNs have the same structure: a black hole, surrounded by a high-speed disc. Gas falls from the disc into the black hole, where it is compressed by intense pressure, heats and emits X-rays. The X-rays and friction heat the disc in turn, producing intense ultraviolet and optical light.

But if AGNs all have the same structure, centred on their central black hole, why do they look so different? This was explained by the 'unification' model of AGNs, put forth in 1984 by Robert Antonucci and Joseph Miller of the University of California, who proposed that the various types (including quasars, radio galaxies and Seyfert galaxies) differ only in the angle at which we happen to view the disc. Surrounding the black hole and the hot inner disc are the cooler outer parts of the disc – an opaque, rotating doughnut ('torus') of dust and thick gas, which has a radius of a few light years. If we happen to view an AGN edge-on, the torus obscures the inner parts, making it a Seyfert galaxy. However, if we see the AGN from a different angle, with the dust ring framing the nucleus, the inner parts of the AGN are revealed, including the blaze of light from near the nucleus, and we see a quasar.

In 1995–99 Cambridge astronomer Andy Fabian used the Japanese satellite Asca to map the X-ray emission from the disc orbiting near the black hole in the AGN galaxy MCG-6-30-15. He found that the X-ray spectral lines from the disc had been bent into a unique shape, because of the combination of several effects of Special and General Relativity, generated by the black hole's powerful gravity. Even if black holes are necessarily permanently secrets in themselves, their surroundings give them away.

Fabian's proof of the existence of the accretion disc surrounding a black hole was an inference from subtle reasoning. To observe this phenomenon plainly, a large team led from the Netherlands by the German radio astronomer Heino Falcke in 2017–18 linked together radio telescopes all over the world to produce one, virtual, global-sized radio telescope, called the Event Horizon Telescope (EHT). The EHT created a radio picture that showed the event horizon of the black hole in the galaxy M87. The image was distorted and magnified by the strength of the gravity around the black hole, but showed the shadow of the event horizon within the radio-emitting material (plate XXVIII). The size of the event horizon of M87 is comparable to the size of the Solar System, but, while our Solar System has one solar mass within it (our Sun itself), the event horizon of M87 has 6.5 billion solar masses within. The picture was a turning point in the history of black holes – the moment that astronomers passed from imagining black holes to imaging them. It was remarkable that the first close-up image of a black hole looked so much like a hole that was black!

The Black Hole in our Galaxy
A dormant monster

> 'You would hardly think, at first, that horrid
> monsters lie up there waiting to be discovered by any
> moderately penetrating mind, monsters to which
> those of the oceans bear no sort of comparison.'
> 'What monsters may they be?'
> 'Impersonal monsters, namely, Immensities.'

Thomas Hardy, *Two on a Tower*, 1882

At the centre of our Milky Way galaxy, a cluster of a hundred stars is orbiting a mysterious object at incredibly high speeds. Although astronomers cannot see the object, they know its mass is over 4 million times the mass of the Sun. It seems to be a supermassive black hole that is invisible because it is sleeping.

Karl Jansky's and Grote Reber's maps of the radio sky showed that radio waves came from along the line of the Milky Way with a maximum in the direction of Sagittarius. The centre of the Galaxy lies about 25,000 light years in this direction, but in visible light we can only see as far as perhaps 1,000 light years, because the interstellar space contains a smoke of dust particles blown out by stars. There is typically 1 dust particle per 1,000 cubic metres (say one particle in a cathedral-sized volume), but since interstellar space is so large, the individual dust particles add up to an opaque screen. Radio waves and infrared radiation can, however, penetrate through this smoke, and reveal what lies at the centre of the Galaxy beyond.

At first the discrimination of radio telescopes was not fine enough to show anything other than that there was a strong radio source in Sagittarius. It gathered the name Sagittarius A as the strongest source in that constellation, or Sgr A for short. It soon became clear that Sgr A was complex, and made up of several different things, like nebulae and supernova remnants.

Radio astronomers discovered that there were two strong radio-emitting sources with distinct radio properties within the region, and so Sgr A was divided into Sagittarius A West and Sagittarius A East. Sgr A East is probably a supernova remnant, but Sgr A West was more of a mystery. It is a complicated spiral shape and it coincides with the highest density of stars in the Galaxy. Radio astronomers interpret it as the centre of our Galaxy, and when in 1959 the International Astronomical Union agreed to set up a coordinate system for the Galaxy, Sgr A West was defined as the central point.

Because Sgr A West is so complex, it was difficult to study. In February 1974 Bruce Balick and Robert Brown used the Green Bank 35-kilometre radio-linked interferometer of the National Radio Astronomy Observatory to map Sgr A West and discovered a bright point-like radio source at its centre. They concluded: 'The unusual nature of the sub-arcsecond structure and its positional coincidence with the inner 1-pc [parsec] core of the galactic nucleus strongly suggests that this structure is physically associated with the galactic center (in fact, defines the galactic center).' Robert Brown invented the name Sagittarius A* (pronounced 'Sagittarius A-star' and abbreviated Sgr A*) for the source.

With the position of Sgr A* clarified and with the development of very large telescopes and detectors optimized for use in the infrared spectrum, it became possible to take infrared pictures of the galactic centre. Infrared pictures show dust, stars and gas lying in the central 30-light-year region. Its very centre is surrounded by the so-called Circumnuclear Disc of dust. The Disc surrounds a cluster of stars, and Sgr A* lies at the centre of the Disc and the cluster.

The motion of these bright stars has been studied by teams led by the German astronomer Reinhard Genzel and University of California astronomer Andrea Ghez. They have repeatedly imaged the stars over the past decade with telescopes in the European

Southern Observatory in Chile and the Keck Observatory in Hawaii and been able to see the progressive displacement of the stars even though they are at a distance of 25,000 light years. Both groups use advanced image-stabilization techniques to overcome the wobble of the star images caused by the Earth's moving atmosphere, which blurs ordinary images. The instrument produces the main image by reflecting it off a series of movable mirrors. The instrument senses the distortion of the atmosphere and adjusts the mirrors to compensate. The sharpened images make it possible to see easily the small displacements of the stars from year to year.

These astronomers have discovered that the stars in the cluster are speeding round Sgr A*. Every fifteen years, one of them, the star known as S2, approaches as close as 120 times the Earth–Sun distance of the black hole – the distance of the furthest small planets in the Solar System. The motions of the stars make it possible to estimate the mass in the cluster. It is 4,100,000 times the mass of the Sun. This cannot be the mass of the cluster itself, which consists of something like a hundred stars. The vast majority of it must be accounted for by Sgr A* itself. And the close approach of S2 means that these millions of solar masses are packed into something the size of our Solar System, which contains just one star. The only object that could be so dense is a supermassive black hole.

If we compare the black hole in our Galaxy with a typical quasar or other Active Galactic Nucleus, it does not rank very high in brightness. It does not rival, say, 3C 273 or M89. One reason for this is that the black hole in our Galaxy is not at the top end of the range of masses for supermassive black holes. Another is that not much matter falls onto it. Collisions between clouds of material in the Circumnuclear Disc causes them to fall in towards Sgr A*, but only a small trickle of material dribbles onto the black hole. Our Galaxy's black hole is thus dormant, its signs of life faint snores when compared to the terrible rages of 3C 273 or M89.

Gamma-Ray Bursters
The biggest bangs since the Big Bang

> From the conception the increase.
> From the increase the swelling.
> From the swelling the thought.

A Maori song, collected by Richard Taylor, 1850

Gamma-ray bursters are cosmic explosions of extraordinary energy, which were first discovered during the Cold War by US satellites designed to detect Soviet nuclear-weapons tests. The Partial Test Ban Treaty, signed in October 1963, prohibited tests of nuclear weapons in the atmosphere or in space. To verify whether the other side was keeping to the terms of the treaty, the US Air Force launched a series of satellites, named Vela, to look for signs of illegal tests.

The satellites watched for the nuclear effects of the explosions – brief bursts of gamma rays. However, from the launch of the first satellites in 1967, the gamma-ray equipment on the satellites discovered a completely unexpected phenomenon – not bursts from man-made terrestrial explosions, but bursts from natural causes. Each burst lasted only about a second and, as they were being detected on an almost weekly basis, were too frequent to be clandestine nuclear explosions from any country.

As the Vela series of satellites became more sophisticated, they provided more information about the mysterious gamma-ray bursts. Vela 5A and 5B (launched in 1969) and Vela 6A and 6B (1970) were able to characterize the signatures of the bursts, even though they were so brief. In 1972 Ray Klebesadel and Ian Strong of the Los Alamos Scientific Laboratory, New Mexico, studied secret records that revealed the direction from which the gamma rays had originated. They found that on several occasions the same burst had been detected by two satellites. To look for tests on both sides of the Earth, the Vela satellites were operated in pairs, identical satellites on opposite sides of a circular orbit 250,000 kilometres in diameter. The scientists found that the gamma-ray bursts were not quite simultaneous in both satellites; in fact, the light travel time from one spacecraft to another, across the orbital diameter, was about 1 second. The gamma rays, travelling at the speed of light, triggered first one satellite and then the other, thus indicating from which direction they had come.

By 1973 Klebesadel and Strong were able to prove conclusively that the gamma-ray bursts were of cosmic origin. They had noticed that some bursts had been seen by four satellites – both the Vela 5 and the Vela 6 pairs – and therefore could only have come from random areas of the sky, but astronomers debated their origins. Klebesadel and Strong were given permission to publish this discovery as a scientific paper, but they were not allowed to report all the details of the instrumentation and its capability, as that might have made it possible for other countries to tailor illegal weapons tests that could not be detected. Some astronomers complained that they were not privy to all the relevant scientific background, but they had to put up with it.

Scientific satellites were deployed on the problem and gamma-ray bursts were discovered almost daily. The phenomenon was given a new name – a gamma-ray burster. Up to 1997 the best clue to the nature of bursters was that gamma-ray bursts come from all directions equally (isotropy) but this evidence was equivocal. Some astronomers thought that bursters might occupy a region surrounding the Solar System at a distance of up to a light year (comets come from this region). Others thought that bursters might occupy a halo, extended around our Galaxy.

By contrast with these possible 'local' origins, some daring astronomers began to consider that the bursters might be 'cosmological' – distributed among the most distant galaxies, at distances of hundreds of millions of light years. This was daring because, if bursters were really this distant, their energy was enormous.

In 1997 the question of the distance and energy of bursters was answered through a key observation by an Italian–Dutch satellite called BeppoSAX. At 5 a.m. on 28 February, BeppoSAX detected a gamma-ray burst known as GRB 970228 – gamma-ray bursts are numbered by date. The satellite's operations team in Rome rescheduled its observing programme to deploy more accurate X-ray sensitive instruments for follow-up observations only eight hours later. They saw a new X-ray source, which quickly faded. Both the positional coincidence and the variability indicated that the gamma-ray burst and the X-ray emission originated from the same object.

Less than one day after the first detection of the gamma-ray burst, Dutch astronomer Jan van Paradijs used the William Herschel Telescope (WHT) on the island of La Palma, Spain, to discover a faint optical source at the same position in the sky as the X-ray source. The WHT's optical images showed that the optical burst was surrounded by a fuzzy patch. The Keck 10-metre telescope on Mauna Kea, Hawaii, then the largest in the world, and the Hubble Space Telescope revealed that the fuzzy patch was a galaxy at an enormous distance from Earth.

To be able to be detected at such distances, gamma-ray bursters let loose as much energy as a supernova in a few seconds or less. In fact, at least some gamma-ray bursters have subsequently proved to have spectral properties that are just the same as imploding supernovae. For some reason the implosion is 'naked' and the gamma rays get out, rather than being absorbed in stellar debris surrounding the explosion.

But not all gamma-ray bursters are the same. About a third of the bursts are much shorter-lived than the majority, typically less than a second in duration. The favoured theory is that they are the result of the merger of two neutron stars, spiralling together in a double-star system. One such event was seen in 2017 as an accompaniment to a burst of gravitational waves. The signature of the gravitational waves was of the merger of two neutron stars, each between 1 and 2 times the mass of the Sun. A short gamma-ray burst followed 1.7 seconds afterwards, and then an optical source that lasted for a month and an X-ray source that lasted over a year. It was clearly a complicated event, which offers the promise of understanding these short-period bursters. It is rare for bursters to happen in a given galaxy, but, since the gamma-ray bursters are powerful enough to bring the entire Universe of galaxies within view and there are so many galaxies in the Universe, astronomers see one happening practically every day.

The Evolving Universe
The past, the present and the future

Of the forces which are imperceptible forces, none is greater than that of change. All things are ever in the state of change. Therefore the I of the past is no longer the I of today.

Chuang Tzu, fourth-century BCE Taoist philosopher

Is the size of the Universe fixed and unchanging, or is it continually expanding? Has the Universe always existed, or did it have a discrete, explosive beginning in an event called the Big Bang? Throughout the twentieth century, rival camps of physicists and astronomers fought over these questions about the Universe's past, present and future.

In the seventeenth century Isaac Newton applied his theory of gravity to the Universe, on the assumptions that the mass in it was uniform and static. Each mass particle – he thought of them as stars, but we would think of them as galaxies – attracted all the others, and Newton realized that the Universe would therefore logically collapse in on itself. Manifestly it has not done so.

Newton never resolved this difficulty. When in 1915 Albert Einstein discovered a new formulation of the theory of gravity, which came to be called the General Relativity, he also tried to apply it to the entire Universe, with the same result as Newton – a static Universe of galaxies was fundamentally unstable. His solution was to invent the so-called Cosmological Constant, which acted like a repulsive force that held the Universe up.

That very year, Willem de Sitter discovered in the mathematics of General Relativity something that Einstein had overlooked: the Universe need not be static, but could be expanding. Around 1927 the Belgian astronomer Georges Lemaître, visualizing the start of the Universe as an exploding atom, proposed theories of the expanding

Universe that were valid without need for Einstein's Cosmological Constant. In 1929 the US astronomer Edwin Hubble used the 100-inch Mount Wilson telescope to map the motions of the galaxies, and discovered Hubble's Law, in which distant galaxies are moving away from us at a rate that is proportional to their distance. The interpretation is that the Universe is expanding. In his 1927 paper Lemaître had implicitly predicted a linear velocity–distance relation of this kind.

Lemaître identified the explosion of the original atom as the moment of the creation of the Universe – the Big Bang. There were some astronomers in the 1950s who had philosophical or religious objections to this, among them Cambridge mathematicians Hermann Bondi and Thomas Gold, who wished to construct a theory in which there was no question of how the Universe originated. They were joined by the physicist Fred Hoyle. After the failure of George Gamow's theory of the origin of the elements in the hot Big Bang, Hoyle had been successful in formulating a theory of the origin of the chemical elements in stars. If there was no Big Bang then clearly there must be other places where they formed, and Hoyle thought he could say what those places were.

The outcome of these discussions was the Steady State theory, which holds that the Universe has always been the same, rather than evolving. Since Hubble's Law showed that it was expanding, getting ever bigger, then something must be filling the space thus created at just the right rate. Fred Hoyle invented the idea of the Continuous Creation of hydrogen, spontaneously in the gaps that developed between the galaxies. In his arguments for the Steady State/Continuous Creation theory, Hoyle invented the derisory term 'Big Bang' to describe the origin of the Universe in the rival theory that it was expanding expontentially – to his surprise, the term was taken up without rancour.

The main difference between the two contrasting theories was that in the one case the Universe had not changed, and in the other it had. In the Big Bang theory, the fundamental difference between the past and the present is that in the past the Universe was denser, with galaxies closer together. Looking back into the past is the equivalent, for astronomers, of looking into the distance: light from a galaxy and its neighbours travels at the speed of light so it carries an image of the galaxy and its neighbours as they were when the light left. The distances to which optical telescopes could see were not so great that this could be investigated, but radio waves travel at the same speed as light, so radio astronomy does just as well. In the 1950s radio telescopes were discovering thousands of radio galaxies at large distances, and the question could be addressed.

Several groups of radio astronomers began to address this issue at about the same time, and gathered themselves in two camps, one led by Martin Ryle at the University of Cambridge and the other a loose alliance of research groups in Australia at the CSIRO Division of Radiophysics and the University of Sydney led by John Pawsey, Bernard Mills and Bruce Slee.

The cosmological question boiled down to arcane arguments about what was called 'log N-log S', which was a relationship that expressed the number, N, of radio galaxies of a given brightness, S. The fainter galaxies are in general further away. Radio waves have travelled a long time to get from the further galaxies and we view them as they were in the past. If there are more of the fainter further galaxies than you would have expected, as indicated by the log N-log S distribution, this must mean that the Universe was denser in the past than it is now, and the Universe must have evolved.

The log N-log S debates got bitter. After the first experimental surveys, Ryle's group published in 1955 a catalogue of nearly 2,000 radio sources, called 2C. It clearly showed a large overabundance of faint radio sources. In Australia, Bernard Mills had just started making a survey with a radio telescope built on lines that were different from the Cambridge ones. Fred Hoyle wrote a worried letter to Mills asking if his results confirmed Ryle's. Even early on they did not confirm Ryle's results. By 1955 the Australian radio-source survey was showing a slight excess of faint sources, but nothing like the excess found by Ryle. The Australians privately expressed reservations about the 2C catalogue, and suggested that most faint sources in the 2C catalogue were spurious instrumental effects.

Ryle publicly ignored the criticisms, and relations between the two groups soured. In 1957 Mills and Slee published a catalogue that amounted to a devastating criticism of 2C: '...there is a striking disagreement between the two catalogues...discrepancies, in the main, reflect errors in the Cambridge catalogue and accordingly deductions of cosmological interest derived from its analysis are without foundation...'. Hoyle was relieved and continued to develop his Steady State theory.

Eventually the Cambridge radio astronomers came to realize the validity of the criticisms and redesigned their equipment and analysis, producing the 3C catalogue in 1959. Ryle was awarded the Nobel Prize in 1974 for 'pioneering research in radio astrophysics...for his observations and inventions, in particular of the aperture synthesis technique'. Third parties, like John Bolton, working with a group at Owens Valley, California, joined the debate and pressed the observations towards their ultimate conclusion, which came when Ryle replaced his 3C catalogue with a revision, 4C. This came much more strongly to the same answer, that the Universe was not the same in the past as now: it was not in steady state, but had changed.

Ryle produced his 4C catalogue in 1965, and invited Hoyle to the press conference at which he would release his results. Hoyle, not informed about what Ryle would say, was sat on the stage and made to listen to Ryle describing what historian of astronomy Simon Mitton has termed 'the inadequate 1C, the hopeless 2C, the ambiguous 3C and the perfect 4C'. 4C proved that the Steady State theory was wrong – there was indeed an excess of faint radio sources at vast distances and look-back times. Hoyle was thus humiliated in public by the stage-managed presentation of Ryle's new results, but astronomers accepted the outcome: the Steady State theory, at least in its original form, was dead. Radio astronomers had discovered that the Universe has evolved.

Cosmic Microwave Background
The after-glow of the Big Bang

Philosophically, I liked the steady-state cosmology.
So I thought that we should report our results as a
simple measurement; the measurement might be
true after the cosmology was no longer true!

Robert Wilson, 'Discovery of the Microwave Background',
in *Modern Cosmology in Retrospect*, 1990

The Cosmic Microwave Background is the black-body (thermal electromagnetic) radiation left over from the fireball of the Big Bang. Between 1946 and 1965 three groups of people were tracking it down, none of them aware of the efforts of the others, all of them starting from different theoretical standpoints, and none of them realizing that its existence had first been implied by a curious, unexplained measurement made in 1938.

In Russia, George Gamow, Ralph Alpher and Robert Herman predicted the existence of the Cosmic Microwave Background (CMB) in 1947 from their theories about the formation of the elements in the Big Bang. In the first seconds of the Big Bang, the heated, expanded material cooled to a temperature of about a billion degrees. Their idea was that during this time elements were created in stages from the simplest element – hydrogen – by the successive addition of neutrons, one at a time, to make heavier and heavier nuclei.

The theory failed because it could not create elements heavier than lithium. However, their analysis of the conditions of the matter in the early stages of the Big Bang became very important: they realized that the early Universe contained hot radiation. The radiation cooled as the Universe expanded, but became frozen when the Universe became transparent, about 400,000 years after the Big Bang, and the expanding matter could not alter the radiation's characteristics.

From then until now, the radiation propagated unhindered through the Universe, to become the CMB. In 1948 Gamow calculated the temperature of the radiation at 10 K (degrees above absolute zero) – Alpher and Herman got 5 K. These independent calculations were remarkably consistent, considering the uncertainties.

Meanwhile, in the USA, Robert Dicke at Princeton University was also studying the formation of the elements in the Universe. He thought the Universe was oscillating: expanding and then collapsing periodically, to what he called a Big Crunch. Each cycle destroyed the elements made in the previous period, evaporating them when the temperature reached 10 billion degrees. The radiation formed in this phase then cooled and became frozen, as in Gamow's theory. Dicke's colleague, Jim Peebles, calculated the temperature of the radiation. In 1964, the two of them and David Wilkinson set out to detect it by building a so-called 'Dicke radiometer', a type of detector that the Princeton researchers had earlier been using to detect the heat of the Moon. In 1946 it had already been used to set an upper limit to the background radiation of 20 K.

While the Princeton detector was being built, Arno Penzias and Robert Woodrow Wilson were trying to identify systematically all the sources of noise in a very sensitive receiver-antenna combination at Holmdel, New Jersey, being used by Bell Labs for early communication satellite experiments (bouncing radio waves off the Echo satellites to reflect radio signals from one continent to another). They used a giant antenna in the shape of a huge horn that could be rotated up and down around a horizontal axis, and around the horizon on a circular ground track. Radio waves that entered the horn were reflected along the axis and through a hole where they could be detected by equipment in a cabin. They were able to eliminate or measure all the instrumental effects, including the effect of droppings from two pigeons that had decided to nest in the horn, but they always saw excess noise, for which they could not account.

Penzias and Wilson decided to investigate whether the radiation had an astronomical origin. It seemed to come equally from all parts of the sky. Meanwhile, at Johns Hopkins University, Peebles gave a lecture mentioning Dicke's idea on the cosmic fireball, and his remark found its way to Penzias. In 1965 the Holmdel and Princeton groups got together and decided that what Penzias and Wilson had detected was the Cosmic Microwave Background radiation. They measured its temperature at 3 K, near enough to the early estimates by Gamow, Alpher and Herman. The discovery was published and became a sensation, because it was evidence against the Steady State theory of the Universe. Penzias and Wilson were awarded the Nobel Prize in 1978 'for their discovery of cosmic microwave background radiation'.

The 3 K radiation had in fact already been discovered by Canadian astronomer Andrew McKellar in 1941, but he had not fully realized the implication of what he saw. He found from measurements of spectral lines in interstellar molecules called cyanogen (CN) made by S. W. Adams in 1938 that the molecules of interstellar space were warmed to about 2.3 K. The CN molecules are interstellar thermometers and the effect was clear, but the source of the warmth was only successfully identified by N. J. Woolf and George Field after the discovery of the CMB had been announced. The latest measurement of the temperature of the CMB is 2.7255 ± 0.0006 K, the most accurately measured temperature of a black-body ever.

Several measurements confirmed that the CMB is basically isothermal and isotropic. Isothermal means that the radiation had almost the same temperature everywhere, and therefore must have had a single origin, such as an explosion. If it originated from, say, something happening in galaxies, its properties would alter somehow with the age, or type, or distribution of galaxies. Isotropic means that the CMB looks the same in every direction. This suggests that the CMB is a property of the Universe as a whole, because there is no reason to think that, by and large, any part of the Universe is different from any other. However, isotropy and isothermality are bound to break down at some level, the closer you look at the CMB and the closer you peer into the details of its origin: the CMB may have originated in one explosion, but the explosion must have had some structure.

After some previous unsuccessful searches for irregularities in the brightness of the CMB, NASA's Cosmic Background Explorer (COBE) spacecraft was launched in 1989, carrying an instrument to search for anisotropies. The spacecraft repeatedly scanned the sky, measuring the intensity and the temperature of the CMB everywhere, looking for differences from place to place. The COBE team, led by Berkeley physicist George Smoot and NASA astrophysicist John Mather, discovered that that the CMB had 'shady patches' at the level of 40 parts in a million (much more uniform than the most perfect white paper), indicating minute variations in temperature (plate XVI). In 1992 Stephen Hawking described this as the 'discovery of the century, if not of all time'.

These anisotropies formed when gravity acted on the minuscule fluctuations in density caused by quantum mechanics during the first moments of the Big Bang. They developed into the major structures in the Universe. Eventually these structures made possible galaxies, stars, planets, and you and me, and the characteristics of the structures make it possible to calculate very precisely some of the properties of the Universe, like its age, size and density.

Smoot and Mather shared the Nobel Prize for the discovery in 2006. The patches have been mapped in greater detail by the Wilkinson Microwave Anisotropy Probe (WMAP), launched in June 2001. Definitive measurements, which achieved the ultimate limit of accuracy at which the fluctuations can be measured (because of complicating interference from other astronomical radiations) were achieved by ESA's Planck satellite, launched in 2009.

Gravitational Waves
Whispers of black holes, neutron stars and the Big Bang

**The weakness of the gravitational interaction makes it
unlikely that gravitational radiation will ever be observed.**

F. A. E. Pirani, *Gravitational radiation: an introduction to current
research*, 1962

When the mass distribution of an object changes, the gravity around
it changes, sending ripples through space at the speed of light. These
ripples are called 'gravitational waves'. At one time, even when Pirani
was writing as recently as the 1960s, it was thought that the effects
were too subtle to be detected, but with new technology astronomers
have detected gravitational waves from merging black holes, and
are now trying to develop space technology to detect gravitational
waves generated split seconds after the Big Bang.

Gravitational waves are a feature of Einstein's Theory of General
Relativity. They are emitted from revolving binary stars, rotating
non-spherical stars and collapsing stars (if they do not collapse
straight down, but splatter). Indeed, they are emitted from any
event in which the distribution of mass changes. A car driving by
gives out gravitational waves, but of course its mass is small, so the
gravitational waves are very weak.

Even from something the mass of a star, gravitational waves are
weak. It was only in 2015 that gravitational waves were first detected,
generated in a far-distant galaxy by the merger of two black holes.
This event briefly radiated more energy than the entire luminous
Universe; however, except for two of the most exquisitely sensitive
sets of equipment, no one noticed this event, because the gravitational waves pass right through the Earth, leaving behind almost
no energy – they whisper quietly and few people can hear them.

Although gravitational waves are themselves a recent discovery, their effect was discovered even as long ago as the 1970s in the case of the binary pulsars PSR B1913+16 and PSR B1534+12.

The first binary pulsar, B1913+16, was discovered by Russell Hulse and his PhD supervisor Joseph Taylor in 1973 with the Arecibo Observatory's radio telescope in Puerto Rico (both men were awarded the Nobel Prize 1993 'for the discovery of a new type of pulsar, a discovery that has opened up new possibilities for the study of gravitation'). The binary pulsar is a pulsar in a small, high-eccentricity, short-period (in this case eight-hour) orbit around a second neutron star. Its pulses can be tracked precisely by radio telescopes, and arrive early or late according to whether the pulsar is near or far as it revolves in its orbit. It was clear right from the initial discovery that the pulsar would be a precise test of gravity – so precise that radio astronomers could observe the effects of General Relativity. In fact, after only two years, while Hulse was writing up his thesis in 1975, the first effects of General Relativity had been seen in the pulsar timings. By 1980 it was possible for Taylor to see the effect of the gravitational waves that the pulsar emits. The loss of energy from the system, as the gravitational waves radiate away, causes its orbit to shrink by 1.5 centimetres every orbit, so it has shrunk by 500 metres since it was discovered.

Pulsar B1534+12 was discovered in 1990 by Aleksander Wolszczan. It has not been observed for as long as the Hulse–Taylor pulsar, but its pulses are both stronger and narrower, so its orbit is clearer. A 'double pulsar', PSR J0737-3039, was discovered in 2003, in which both neutron stars are pulsars. The extra precision that this enables makes this by far the best pulsar system with which to test General Relativity. In all these cases, it has been possible to measure not gravitational waves themselves, but their effects, and the calculations of General Relativity fit amazingly well, to an accuracy much more precise than 0.1%.

The first successful attempts to detect gravitational waves directly were made in 2015 by detectors in Hanford, Washington, and Livingston, Louisiana, respectively, which together make up the Laser Interferometer Gravitational-Wave Observatory (LIGO), working in conjunction with a similar detector near Pisa, Italy, called VIRGO, with further detectors being developed in Hanover (GEO 600), Germany, and in Tokyo (TAMA 300), Japan. The principle of the instruments is to measure with a laser the distance between two freely hanging pendulum mirrors. The passage of a gravitational wave causes the mirrors to bob like corks on the sea, changing their separation.

In 2015 LIGO saw the first detected event of gravitational waves from two merging black holes. By the end of 2018, ten such phenomena had been observed, together with an eleventh event that was generated by two merging neutron stars. In 2017 the Nobel Prize in Physics was awarded to Rainer Weiss, Kip Thorne and Barry Barish for their role in making the first detections. The first event, GW150914, lasted only two tenths of a second, as the two black holes circled each other in their final orbits, quickly spiralling inwards so that their orbital speed increased in a 'chirp' from 35 revolutions per second to 250 per second. They touched and merged into one black hole, oscillating afterwards for a few hundredths of a second. The two black holes were 35 and 30 times the mass of the Sun, with the outcome being a black hole of a mass 62 times the Sun. The missing 3 solar masses had been converted to the energy of the gravitational waves, through the $E = mc^2$ equation. The merger took place in a galaxy at a distance of about 1.5 billion light years from Earth, so practically the entire Universe lies within sight of the detectors for this kind of event.

Over the succeeding years, LIGO astronomers refined their data analysis capabilities and identified in their archived data a black-hole merger that preceded the first one identified, a gargantuan collision seen on 29 July 2017. Two black holes, one 50 and the second 34 times the mass of our Sun, merged to make a single black hole over 80 times the mass of our Sun. Its galaxy is 9 billion light years away. The 50-solar-mass black hole is the largest of the twenty black holes found so far through gravitational wave observations of black-hole mergers.

The signature of the gravitational waves from the mergers was precisely what had been expected, which made analysis of the observation straightforward – the way to do it had all been worked out before. What was unexpected was that the black holes were so big. Black holes in a binary star system come from supernova explosions of the progenitor stars. Cutting a long story short, astronomers believed until then that only smaller black holes could be the result of a supernova. There must be something that astronomers do not understand – but what? The answer is a discovery yet to be made.

The neutron-star merger event, GW170817, took place in 2017 (plate XXI). The two neutron stars spiralled into each other over a period of nearly two minutes, speeding up from 24 revolutions per second to about 300. The merger seemingly produced a neutron star of up to about 3 solar masses. Considering that neutron stars can survive only if their mass is less than 2 solar masses, it seems that this one was hypermassive and may then have collapsed to a black hole. Black hole mergers produce no light, radio or X-ray energy, but a neutron-star merger splatters about material that picks up energy from the event and radiates in ways that optical, radio and X-ray telescopes can see. The event was seen by seventy observatories world-wide and in space as a brief optical flash lasting a few days, and a burst of X-rays.

The first space-borne gravitational-wave detector, called the Laser Interferometer Space Antenna (LISA), is being developed by the European Space Agency. It is planned as three spacecraft, spaced at 2.5-million-kilometre intervals in an equilateral-triangle formation, following the Earth around in its orbit. It will contain devices to counter the effects of the solar wind as it buffets the spacecraft. The effects of gravitational waves as they pass through the Solar System will be detected by seeing how the spacecraft oscillate. The spacecraft will communicate by lasers that measure how far apart they are. A test vehicle, the LISA Pathfinder, was successfully used to verify some of the engineering techniques in space in 2015. LISA will be the largest man-made construction ever. LISA will be more sensitive than current gravitational-wave detectors partly because it will not be subject to the earth-tremors that shake the ground-based instruments, but also because of its size – the earth-based detectors are only 300 metres to 4 kilometres long.

Gravitational waves have been detected only from binary stars, but it is predicted that they could also come from any spinning neutron star that has a bump on one side. Usually neutron stars have strong gravity and are spherical, but some neutron stars in some binary stars sit under an infalling stream of gas that heats the surface of the neutron star and may cause a hill. Gravitational waves can also come in bursts from supernovae that collapse to a neutron star or to a black hole. Finally, gravitational waves were made in the early history of the Universe and carry a picture of the conditions in the Big Bang as it was 1-million-million-million-millionth of a second old, just as the Cosmic Microwave Background carries a picture of the Universe when it was 400,000 years old. LISA should detect this Gravitational Wave Background.

Darkness at Night
The missing galaxies

> From innumerable stars a limitless sum total of radiations
> should be derived, by which darkness would be abolished
> from our skies and the 'intense inane [void]', glowing with the
> mingled beams of suns individually indistinguishable, would
> bewilder our feeble senses with its momentous splendour.

Agnes Clerke, *The System of the Stars*, 1890

The darkness at night hides surprising secrets. It tells us about our place among the stars, and it shows that the Universe has not existed forever.

The fact that it is dark at night is such a commonplace that it cannot be said to be a discovery. But within the commonplace fact lie surprisingly deep discoveries about our cosmic situation. In the first place, it tells us that there is only one star in our neighbourhood – the Sun that our planet is orbiting. It is rather rare for a star to be single like this – most stars have companions close by, in double or triple stars, or in groups or clusters. In science fiction, scenes may take place on a planet that has two suns, as in the film *Star Wars*, where we see them in the background, both in the sky together. Such a planet must have complex day/night patterns. Or a planet might orbit a star in a cluster of stars, in which case the other stars would be scattered all over the sky and shining brightly. It would never be as dark on such a planet as our night, although there may be a bright/less bright cycle.

These scenes are imagined. But the reality is that we do live in a cluster of stars of sorts – the Galaxy. At night we do see its constituent stars, and they cast a dim light on the surface of the Earth. This is the reason why it is possible to see more on a star-lit night than on a cloudy one, as hunters and soldiers know well.

How bright is it at night? There is an argument that it ought to be as bright at night as by day. The steps to this argument were discovered progressively in the eighteenth and nineteenth centuries, starting after it became clear that the Universe was not contained within crystal spheres, but extensive, with many stars beyond the edge of the Solar System. It might even be infinite in extent, and therefore, assuming that it is isotropic, contain an infinite number of stars. Edmond Halley was one of the first to consider this possibility in papers published in 1721 under the name 'Of the infinity of the sphere of the fix'd stars' and to link the possibility with the question about the light of the night sky. The mathematical argument was fully formulated by the Swiss astronomer Jean-Philippe Loys de Chéseaux in 1744.

Think of an infinite universe of stars, uniformly distributed around the Earth, which we can visualize as a succession of thin spheres of the same thickness, each sphere getting larger and larger like the shells of an onion. The number of stars in each shell is proportional to its volume, and increases in proportion to the square of the radius of each shell. But the light that reaches the Earth from each star in each shell diminishes according to the inverse square law, so the total light at the Earth from each shell is the same. If there is an infinite number of shells, the light from all of them added together is infinite – there would be no dark sky at night.

Chéseaux realized that there is a limitation in this argument – the stars in the nearer shells would obscure some of the stars in the shells behind. The most distant stars would not contribute any light at the Earth, just as, in a forest, no matter in which direction we look, we see a tree trunk, whether nearby or further away. We cannot see any of the trunks of the most distant trees. Substitute 'star surface' for the words 'tree trunk' in this analogy and it becomes clear that based on the argument laid out, no matter where we looked in the night sky our line of sight would end on the surface of a star. The surface brightness of the sky would be the same as the brightness of the surface of an average star, like our Sun. Chéseaux estimated that we could see on average a distance of 3,000 trillion light years, that the number of visible stars would be correspondingly many, many trillions, and that the sky at night would contribute as much light as 90,000 suns. His figures are broadly consistent with modern calculations.

Chéseaux recoiled from these numbers, and the inevitable conclusion that the sky at night would be as bright as the surface of the Sun, which is even brighter than the sky by day. He proposed that the solution to the problem was either that the starry part of the Universe was finite, terminating well before the sky was completely full of stars in no matter which direction you looked, or that starlight was absorbed in space, diminishing the light from the more distant stars. The enormous difference between his theoretical conclusion and our actual experience demonstrates either that the sphere of fixed stars is not infinite, or that something unknown blocks starlight.

The problem of why it is dark at night was revisited seventy years after Chéseaux published his paper by Heinrich Wilhelm Matthias Olbers. Olbers was a doctor in Bremen, Germany, who discovered two of the first asteroids. In 1823 he wrote 'On the transparency of space', in which, without acknowledging Chéseaux's work, he considered Chéseaux's problem and proposed Chéseaux's solution. Olbers' work became well known but Chéseaux's did not, and as a result the problem formulated carefully by Chéseaux is known as Olbers' Paradox.

In fact, one of Chéseaux's solutions to the paradox raises a further problem, discovered by John Herschel in 1848. If starlight is absorbed by something in space, that something must get hotter, until eventually it gets as hot as an average star's surface and gives out as much light as it takes in. Putting something in space to block starlight does not, in the long term, help solve Chéseaux's and Olbers' problem.

In modern times Olbers' Paradox has relevance to cosmology, if you substitute 'galaxies' for 'stars'. In 1960 the Austrian-born

British mathematician Hermann Bondi listed the four major assumptions of Olbers' Paradox as it would be formulated for cosmology: (a) The Universe is uniform throughout space; (b) The Universe is unchanging in time; (c) There are no major systematic motions in space; (d) The laws of physics apply everywhere. With these assumptions about the galaxies in space, Chéseaux's and Olbers' calculation holds together. But modern cosmology offers possible solutions, by attacking assumptions (a), (b) and (c): (a) The Universe is not infinite and therefore not uniform beyond a certain distance; we can see out only as far as it has been possible for light to travel since the Universe formed; (b) In fact, we can see galaxies only out to the distance and time at which they were formed – before that time, and beyond that distance, the Universe was different; (c) Moreover, the Universe is expanding, with the galaxies receding, and light from the more distant galaxies is weakened by the recession.

The last objection is not as important as the other two, which are of course connected. The sky is dark at night because of missing galaxies, not because of missing light from galaxies. The simple fact that the sky is dark at night means the Universe was created a finite time ago, and is limited in dimension.

FUTURE

DISCOVERIES

Dark Matter
A known unknown

> There are known knowns: things we know that we know. There
> are known unknowns: things we know we don't know. But
> there are unknown unknowns: things we do not know we don't
> know. Each year we discover further unknown unknowns.

Donald H. Rumsfeld, US Secretary of Defense, 2002

Astronomers estimate that 80% of the material in the Universe
is a substance called 'dark matter'. It is invisible to all currently
available technology, and nearly everything about it is a secret yet
to be uncovered, leading some scientists to question whether dark
matter exists at all.

Looking out at the broad view of the Universe, astronomers see
mass on a large scale, distributed as stars and gas in galaxies and in
clusters of galaxies. There is also a considerable amount of hydrogen
and helium in intergalactic space in clouds that do not shine and
contain few or no stars – galaxy-sized clouds left over from the Big
Bang that have not turned into galaxies. This material makes its
presence known by absorbing ultraviolet light from distant quasars
– the light from a distant quasar penetrates through the clouds as
if they were pieces of meat skewered on a kebab, and each cloud
leaves a gap in the spectrum of the quasar's light.

There is also a considerable amount of matter in the Universe
that astronomers cannot see, either by the light (or other radiation)
that it emits or through its absorption of light that encounters it.
This mysterious substance is generically called 'dark matter'.

The person who discovered dark matter was the notoriously
abrasive Swiss-born astronomer Fritz Zwicky of the California
Institute of Technology. In the 1930s he set out to provide a com-

plete scheme of every thing in the Universe. Measuring the mass of every known space object was a good place to set the boundaries of such a scheme. In 1933 he estimated the mass of a nearby cluster of galaxies in the constellation Coma by measuring the speeds of motion of its galaxies, each of them pulled in orbit by the mass of the cluster as a whole. They were moving much faster than Zwicky expected, based on his estimate of the mass of all the stars in the cluster as judged by the light they gave off – his estimate was that the Coma Cluster was 400 times the mass of its stars and that $^{399}/_{400}$ths of the mass was therefore, in his words, 'missing'.

Zwicky's discovery of dark matter in clusters of galaxies was at first not followed up by his colleagues. It was an outlandish idea, made even less palatable because he was so difficult to converse with. However, it was confirmed half a lifetime later on a smaller scale, not within clusters of galaxies but within galaxies individually, by Carnegie Institute astronomer Vera Rubin. When she graduated from Vassar College, New York, in 1948, she applied to study for a further degree at Princeton University but was told that the university did not admit women to its astronomy programme, and so she developed her career instead by studying at Georgetown University, where she was supervised by George Gamow. She went on to work at the Carnegie Institute with Kent Ford, who had developed a sensitive spectrograph that could measure the spectrum of faint galaxies. Rubin and Ford used the new spectrograph to determine how spiral galaxies rotate, and pressed their observations further into the outer, fainter regions of the galaxies than had been possible before.

The expectation was that stars in the outer regions of a spiral galaxy would move more slowly than stars in the central regions, because most of the mass of stars is concentrated in the central regions, and stars in the outer regions are a long way away. In just the same way, the more distant planets of the Solar System rotate around the Sun more slowly than the inner planets. But Rubin and Ford discovered that, in fact, stars in the outer regions of each spiral galaxy were moving just as fast as those nearer its centre.

This means that there is unseen matter in each galaxy, even in the outer regions of galaxies where the visible stars are few. There is at least twice, and up to ten times, more mass in a typical spiral galaxy than is accounted for by stars. By 1975 Rubin had become convinced that 'What you see in a spiral galaxy is not what you get.'

Initially she met with the same scepticism that Zwicky had provoked. But the evidence became overwhelming with the construction of radio telescopes like the Westerbork Synthesis Radio Telescope in the Netherlands, which measured the rotation of hydrogen gas in spiral galaxies, which spreads far beyond the starlight, still with the same result.

In 1937 Zwicky had laid out another way to investigate the mass of galaxies. If by chance a massive galaxy lies along our line of sight to a more distant galaxy, then according to Einstein's Theory of General Relativity it acts as a 'gravitational lens', warping the surrounding space to magnify, distort and displace the image of the background galaxy.

Zwicky did not live to see a gravitational lens discovered. The first one was discovered in 1979 by Dennis Walsh, Robert Carswell and Ray Weymann. They were using a small telescope at the Kitt Peak National Observatory in Arizona to identify quasars discovered by the University of Manchester's Jodrell Bank radio telescope, and found a pair of identical quasars right next to each other – in fact, two images of the same quasar produced by a gravitational-lens galaxy. Since then, gravitational-lensed images have been found that have been produced by clusters of galaxies, and have been used to confirm Zwicky's calculations of the clusters' mass. The latest estimate is that 5% of the matter in the Universe is made up of stars, 15% of intergalactic gas clouds and 80% of dark matter. Altogether these forms of matter make up 32% of the energy in the Universe, with 68% being dark energy.

The composition of dark matter is unknown. It might be some sort of unknown massive elementary particle (different models being known under various names, such as axions or 'WIMPs' – Weakly

Interacting Massive Particles). Laboratory searches for them are under way, and might produce the discovery of the twenty-first century. If dark matter is indeed found in the laboratory, this will be a case like helium, where a fundamental constituent of matter has been discovered in the cosmos before being identified on Earth. However, given that dark matter has not been identified, some astronomers speculate that there is something wrong with our theories of gravity, or that dark matter is ordinary matter in some hard-to-see form. At least we know there is something to explain, somewhere between a known unknown and an unknown unknown.

Dark Energy
On the threshold of a profound discovery

> Much later, when I was discussing cosmological problems
> with Einstein, he remarked that the introduction of the
> cosmological [constant] was the biggest blunder of his life.

George Gamow, *My World Line*, 1970

Imagine taking a region of space and removing all matter and radiation from it until the area is completely empty, much more so than ordinary interstellar space. The result is a 'vacuum'. This vacuum has effects that physicists and astronomers call 'dark energy'. Dark energy is thought to account for nearly ¾ of the energy in the Universe, but, like dark matter, we have not yet discovered a way to see dark energy.

To common sense, the vacuum of space is nothing. To a scientist, the vacuum is not nothing; it is a physical state and it has an energy. In the absence of gravity, there is no way of measuring the energy of a state on an absolute scale; the best we can do is to compare energy differences. The vacuum energy itself would be arbitrary. According to the theory of General Relativity, however, any form of energy has a gravitational effect, so the vacuum energy might be a crucial ingredient in the evolution of the Universe. Colloquially, the vacuum energy is known as 'dark energy.'

Dark energy is as important to the dynamical history of the Universe as dark matter. The Universe is expanding, and it is expanding nearly freely. The galaxies and dark matter in the Universe mutually pull one another, a process that should slow the expansion down. This has been checked by the Hubble Space Telescope. In looking out into the distant Universe, the HST is looking back in time, because light travels at a finite speed and carries pictures of the Universe from far away and in the past to here and now. So the furthest galaxies, representing the earlier Universe, should be moving more quickly than nearby ones. In 1998–99 astronomers from the Supernova Cosmology Project and the High-Z Supernova Search Team used supernovae observed with the HST to check the distances of distant galaxies and the largest ground-based optical telescopes to check how fast they were moving. The two teams discovered the reverse of what was expected. The expansion of the Universe is speeding up, not slowing down. There is some progressive input of energy into the Universe. It goes under the name of 'dark energy', and its nature is a mystery.

Curiously, dark energy has similar effects to a concept hypothesized a century ago by Albert Einstein. Einstein formulated his Theory of General Relativity before the discovery that the Universe is expanding. He used it to develop a theory of a static Universe. The force of gravity is what attracts all galaxies together. If gravity was the only force there was, it would be impossible for the Universe to be static – Einstein needed something to stop the galaxies falling together. Quite arbitrarily, he added a term to his equations called the 'Cosmological Constant', an opposing force cancelling out the effects of gravity, symbolized by the Greek letter Λ (lambda). When the expansion of the Universe was discovered, Einstein retracted the concept as unnecessary and regarded introducing it as a blunder.

As history has turned out, Einstein was ahead of his time by inventing Λ to provide a theoretical solution to a problem that did not then exist. Quite what the theoretical solution really means is still not clear, but some of its consequences are. A cosmological constant has the tendency to cause galaxies to accelerate away from us. In a Universe with both matter and vacuum energy, there is a competition between the tendency of Λ to cause acceleration and the tendency of matter to cause deceleration. This has a big effect on the 'formation of structure', the name that astronomers give to the way that the earliest irregularities in the material of the Big Bang grew. The denser bits drew in surrounding matter and grew to intergalactic-size clouds, which nucleated into clusters of galaxies, in which condensed stars and planets and, indeed, people. The balance between gravitation and Λ controlled the way our Universe turned out.

In a major calculation called the Millennium Simulation (plate XIX), astronomers of the University of Durham and the Max Planck Institute for Astrophysics in Garching, Germany, showed how the formation of a cluster of galaxies starts weakly, because the fluctuations in Big Bang material are not very pronounced. Gravity, principally from dark matter, draws in surrounding material, but then the release of dark energy tends to stabilize the infall, which peters out as the galaxies orbit one another. The Millennium Simulation uses theories of cosmology, hydrodynamics and General Relativity. The output is a sample of the distribution of galaxies, and the actual distribution of galaxies in the Universe can be compared with the simulation. The astronomers can fine-tune the output of the simulation by altering the amounts of presumed dark matter and dark energy so that the calculated and real distributions match. The evidence suggests that the Universe has something like 32% of its density in matter and 68% of its density in dark energy. This indicates the scale of the problem – it is fundamental to physics. As with the problem of dark matter, the problem of dark energy may indicate that there is something wrong with our theories of gravity.

Life in the Universe
Are we alone?

> My Reason is convinc'd, said the Marquiese, but my
> Fancy is confounded with the infinite Number of living
> Creatures, that are in the Planets; and my thoughts
> are strangely embarras'd with the variety that one
> must of Necessity imagine to be amongst them.
>
> Bernard le Bovier de Fontenelle, *A Discovery of New Worlds*,
> translated by Aphra Behn, 1688

No one knows whether astronomers will ever discover life on other planets. If they do, it will prove that we are not alone in the Universe. Most people seem to like this idea, but the Universe has yet to reveal its secrets on the matter.

Life is based on carbon. Carbon is made in stars and is everywhere in the Universe. It is the only element that makes complex 'organic' molecules that form a rich 'vocabulary' that can be used in the 'script' of life. Radio astronomers have discovered about 150 sorts of organic molecules in interstellar space. Comets and asteroids have organic 'tars' and 'crusts' on their surface.

In 1969 a meteorite broke up over Murchison, near Melbourne, Australia, and pieces fell across the town. There were many eyewitnesses to the fall (church-goers on a Sunday morning), who promptly collected about 80 kilograms of fragments. The meteorite was found to contain more than ninety amino acids, a variety of organic compounds essential for life, as well as evidence that they were made in space, rather than being terrestrial contaminants. The implication of all this is that basic biochemicals are made by natural processes in the Universe, and move from place to place.

The French chemists Jean-Baptiste Biot and Louis Pasteur discovered in the early half of the nineteenth century that certain biochemical molecules have a property called 'chirality': they occur in both 'left-handed' and 'right-handed' forms, only one of which is used by life. Thalidomide is a notorious example, in which one chirality, used as a drug, is benign and the other actually very harmful, causing birth defects. Inorganic chemical processes usually produce equal amounts of both chiralities. But if you shine strongly polarized light into a 'soup' in which there are equal numbers of left- and right-handed molecules, some can switch from one sort to the other, and the soup develops an excess of one form of chirality, and may favour biochemical rather than inorganic chemical reactions. Some astronomers think that chirality can be generated in the organic molecules on comets when they pass a source of strongly polarized light, such as a pulsar, transforming them into 'seeds for life'.

Astronomers discovered the processes that delivered the chiral chemicals to the Earth, but geologists identified the 'fire' that simmered them into life. Energy is required to fuel the chemical reactions that change the pre-biotic chemicals into more complex life-forming molecules. On Earth, sunlight is the usual energy source – plants photosynthesize chemicals to create cells – but it is not the only source. 'Black smokers' are geothermal vents in the deepest, darkest parts of the ocean, which are colonized by oceanic life that feeds on geothermal energy rather than sunlight. The microbial forms of life that flourish in such extreme environments are called archaea, and they are the oldest living things on Earth. In fact life seems to have started under the oceans, and only later evolved to harvest energy from sunlight on the surface. Some regions of the satellites of Jupiter and Saturn – Europa, Callisto, Ganymede and Enceladus – mimic this environment of water warmed by internal energy sources, and are regarded as potential sites in which to search for extraterrestrial life.

Biochemistry needs a medium in which to operate: water. Although water is made of simple elements (hydrogen and oxygen) that are abundant everywhere in the Universe, liquid water can only survive on the surface of a planet in the so-called 'Goldilocks Zone' of a planetary system: that is, where the temperature is neither so cold so that water freezes, nor so hot so that it evaporates, but just right. This is mainly a matter of the planet being at the right distance from its sun, but can be helped or hindered by any greenhouse effect in the planet's atmosphere. It is also possible for a planet outside the Goldilocks Zone to have liquid water because of geothermal heating, as on Mars or Europa.

The idea that organic chemicals developed in space and were brought to Earth by cosmic processes, to evolve in the oceans, was first proposed by Isaac Newton and given modern form in 1908–11 by the father-and-son team Thomas C. Chamberlin and Rollin T. Chamberlin. The Chamberlins suggested that 'planetesimals,' the small bodies that merged into the planets, were the source of organic material from which life evolved, creating organic compounds as they collided and accreted with each other. The basic ideas of the theory were revived by Soviet astrobiologist Alexander Oparin in 1938, and reformulated with the aid of modern scientific knowledge by Joan Oró of the University of Texas and Chris Chyba, director of the SETI Institute (Institute for the Search for Extraterrestrial Intelligence) and a student of Carl Sagan.

Another possible scenario for the early formation of life on Earth and elsewhere was discovered in the laboratory. In 1953 Stanley Miller, a PhD student at the University of Chicago, put simple organic chemicals into a flask of water from which oxygen gas had been removed, and then passed an electric spark in through the vapour. Miller did not have deep-sea geothermal vents in mind; he was guided by the idea that electric discharges (lightning) created biological molecules in the atmosphere. Miller found amino acids in the black sludge produced by his experiment. No one knows precisely how life first formed on Earth, but hundreds of similar experiments have since been carried out with a variety of chemicals and energy sources, showing similar results.

Scientists believe that the right chemical ingredients, a source of energy and the presence of water as a solvent at some point combined to transform inanimate organic chemicals into life on Earth, perhaps also on Mars and Europa. In 1996 David McKay and his co-workers discovered what appeared to be traces of bacteria in the Martian meteorite ALH84001, which was found in Antarctica. The meteorite had been knocked off the surface of Mars by an asteroid impact 16 million years ago, and after a period in orbit round the Solar System, fell to Earth 12,000 years ago. It has curious surface shapes that look like fossil bacteria, and contains molecules that were clearly made in liquid water, as well as mineral grains and molecules that are similar to some made organically here on Earth. McKay cautioned that 'none of these observations is in itself conclusive for the existence of past life', but asserted that 'when they are considered collectively, particularly in view of their spatial association, we conclude that they are evidence for primitive life on Mars'. This is a strong claim that needs stronger evidence.

What of life in the Universe beyond the Solar System? There are few, if any, Earth-like planets among the thousands of exoplanets identified by our current technology, but stars are numerous, planets orbit around many of them, and there has been a lot of time for complex life to develop. So why are we not in contact with intelligent extraterrestrials from all these suitable planets? This contradiction is called the Fermi Paradox after the Italian-American physicist Enrico Fermi, who, speaking of extraterrestrials in 1950, asked 'Where are they all?'

Fermi's question was posed expecting the answer that there have been no interstellar visitors, but there has been speculation about an object that, in 2017, visited the Solar System from interstellar space, passing within the orbit of Mercury. Named 'Oumuamua, Hawaiian for 'first messenger from afar', it was our first known interstellar visitor, albeit an inanimate one. Probably a comet-like object from another planetary system, from its variability as it reflected sunlight it proved to be unusually long and thin, and its orbit was not exactly the orbit that it would have followed under the force of gravity alone. The extra force seems to have come from the push of solar radiation on the comet. However, the characteristics of the push suggest that the comet had the structure of an artificially constructed solar sail. This led some to claim that 'Oumuamua was not a comet but an interstellar spaceship, indicating that intelligent extraterrestrials do live beyond the Solar System.

Although we have no firm evidence for it, we can assume that many places in the Universe have the right conditions and ingredients for life. In these circumstances, single-celled life seems to evolve fairly quickly: the oldest fossils of primitive life on Earth are stromatolites, mats of blue-green algae (cyanobacteria) fossilized in rocks as much as 3.5 billion years old. The algae thus represent what happened after the first 20% of the timeline of the history of the Earth. Simple life may well be abundant in the Universe.

However, multicellular life forms and advanced intelligence need – crucially – time to evolve, and time of the right quality may be a much rarer commodity in the Universe than the right conditions and ingredients for single-celled life. Primates have evolved only in the last 85 million years, hominids less than 20 million years ago, and the earliest tool-making hominids 2 million years ago – less than 0.1% of the age of the Earth. Judging by the one example that we know, evolution needs to try out lots of experiments before it can come up with complex life, and requires a platform that is relatively stable for billions of years by way of a laboratory – a planet where temperature and climate change little, and where any sudden change is not too severe. The Earth has had several lucky escapes from catastrophic events. A freak accident (the creation of the Moon) gave the Earth an extra-large iron core and therefore a strong magnetosphere that defends it against the scouring and irradiating effects of the solar wind. The inner Solar System has also survived the migration of Jupiter-sized planets that rampage through other planetary systems.

Your planet has survived all this and produced you. Other planets may not be so lucky, stable or fertile, so while simple life forms may be common in the Universe, complex, and therefore intelligent, life may be rather rare. The distances between the planets that have intelligent life, if there are indeed others, may consequently be so great that it is almost impossible for them to communicate with each other. This may be the explanation for the Fermi Paradox. We may like to take comfort in thinking that there is life elsewhere in the Universe, but we also have to face the possibility that, at least in practical terms, we are indeed alone.

Glossary

accretion The process by which an astronomical body increases in size, by gathering particles of matter to itself, either by means of its gravitational pull or by the adherence of particles with which it collides. The additional matter orbits the main body in a flat 'accretion disc'.

active galaxy A **galaxy** with a powerful and energetic nucleus (that is, an 'active galactic nucleus' or AGN).

Algol paradox (In a **binary star**, such as the prototype, Algol) The anomaly that the less massive star is the more advanced in its evolution.

ansae (Latin: 'handles') Ear-like protrusions from a celestial body.

aperture synthesis interferometry A radio-astronomy technique in which a line of stationary radio telescopes depends on the rotation of the Earth to gather radio waves in a way that simulates the activity of a much larger radio telescope.

asteroid A minor **planet**, of a size in the range of 1 metre to 1,000 kilometres, typically orbiting in the Solar System between Mars and Jupiter. An asteroid is not a **dwarf planet**, a **comet**, a **meteoroid**, or a **moon** or **satellite**. See also **asteroid belt** and **Kuiper belt**. Synonymous with **minor planet**.

asteroid belt The main band of **asteroids** of the inner Solar System, lying between the orbits of Mars and Jupiter; compare **Kuiper belt**.

aurora Light produced by atoms and molecules in the atmosphere and caused by the impact of ionized particles accelerated in a magnetic field. The term is applied especially to the Earth's Northern and Southern Lights.

Becklin–Neugebauer Object (BN Object) A source of intense infrared radiation found in the Orion Nebula and thought to be a newly formed star; it is named after the two astronomers who first observed it.

Big Bang The dense, high-energy, explosive event that occurred at the beginning of the Universe.

Big Crunch The hypothetical high-energy event that will occur if the Universe eventually implodes.

binary In astronomy, a system composed of two bodies in orbit around each other: for example, a binary planet, binary pulsar or **binary star**.

binary star A **double star**, particularly a close pair.

bipolar nebula A symmetrical **nebula** that has two lobes, pointing in opposite directions; often, but not exclusively, a **planetary nebula** of that shape.

black hole An astronomical body that is both small and massive, thus exerting such a strong force of gravity that no light or other radiation can leave the surface.

blink comparator A viewing device used in astronomy to compare two photographs: the viewing optics switch ('blink') quickly between the two images to reveal changes, such as the successive positions of a moving star or planet, or the different appearance of a **variable star**.

Bode's Law (Titius–Bode Law) The law, popularized by Johann Bode (first formulated by Johann Titius), by means of which the distances of the planets from the Sun can be calculated.

caldera (Spanish: 'cauldron') A volcanic **crater**, created when magma sinks from the subsurface, which weakens the surface so that it can no longer support the weight of the material above and the volcano collapses inwards.

canals of Mars (from Italian 'canali') A network of long, straight markings apparently covering the surface of Mars; thought originally to be irrigation canals, they are, in fact, an optical illusion.

catastrophism The theory that geological features are caused by unpredictable, large-scale events (catastrophes) such as floods or meteor impacts. Compare **gradualism**.

celestial sphere (In Ptolemaic astronomy) The imaginary sphere centred on the Earth (or the observer); the fixed stars appear to lie on its surface.

Cepheid variable star A pulsating **variable star** that varies in brightness in a regular cycle, as in the case of the star δ Cephei.

Chandrasekhar mass The maximum possible mass of a **white dwarf** before it collapses under its own weight (to become a **black hole**); by extension, the maximum mass of a **neutron star**.

chaos A mathematical effect: the property of certain equations that predict future behaviour (such as the weather or planetary positions), which are so sensitive to the initial conditions that even the slightest change in the starting point eventually produces a completely different outcome.

chirality Asymmetry of the type in which the mirror image of an object cannot be superimposed on the object itself, no matter how the image is rotated. It is a property of the human hand, and of some of the molecules that are key to biochemistry.

chondrite A type of **meteorite** made of 'chondrules' (near-spherical globules) and other material, which have not been subject to melting, as has happened, for example, in the interior of a planet.

chromosphere (from Greek: 'colour' and 'sphere') The pink-coloured lower atmosphere of the Sun, visible at a solar eclipse; it extends to a height of about 2,000 kilometres and has a temperature of up to 20,000 K.

circum-nuclear disc The outer part of an **accretion** disc in orbit around the **nucleus** of an **active galaxy**.

CNO cycle The nuclear fusion cycle in massive stars, in which carbon, nitrogen and oxygen nuclei progressively combine with protons and eject alpha particles, thus turning hydrogen into helium.

coma (Latin: 'hair') The 'atmosphere' surrounding the solid nucleus of a **comet**; viewed through a telescope, the coma makes the comet look fuzzy, as if it has hair.

comet A small body in orbit in the Solar System; a comet resembles an **asteroid** but is made of icy material, the surface of which melts, surrounding the nucleus with a **coma** and producing a **tail**.

continuous creation A (no longer widely accepted) theory that material is continuously produced in space at such a rate as to ensure that the density of matter in the Universe is constant, even though space is expanding.

Copernican theory The **heliocentric theory**, attributed to Nicolaus Copernicus.

core The innermost region of a **planet**, usually a dense sphere. The core of a terrestrial planet is surrounded by the **mantle**, or outer solid layers.

Coriolis force The apparent force (named after Gaspard-Gustave de Coriolis) that deflects objects moving on a rotating frame. Gives rise to the Coriolis effect – for example, in the Northern Hemisphere, winds are deflected to the right because the Earth is rotating.

corona (Latin: 'crown') The atmosphere of the Sun above the **chromosphere**, revealed in a solar eclipse.

cosmic fireball The extremely intense radiation that occurred at the **Big Bang**, the relic of which is the **Cosmic Microwave Background**.

Cosmic Microwave Background (CMB; cosmic background) An almost uniform source of infrared and microwave radiation, of cosmic origin (see **cosmic fireball**).

cosmic rays High-energy particles from the Sun and the Galaxy that permeate outer space.

Cosmological Constant A term added by Einstein to his General Relativity equations describing the repulsive force between galaxies in the Universe to counter the gravitational force and to prevent the Universe from collapsing.

crater A hole in the ground caused by an explosion or impact.

cubewano A type of object in the **Kuiper belt** that orbits undisturbed beyond Neptune, in the same plane as the major planets.

dark ages In the early history of the Universe, the period after the **Big Bang** but before galaxies became visible.

dark energy (vacuum energy) A hypothetical constituent of the Universe, which, released gradually into space, causes the expansion of the Universe to accelerate.

dark matter A hypothetical, unseen constituent of the Universe, which produces a gravitational attraction on galaxies and stars.

dark nebula A **nebula** of **interstellar dust** that makes its presence known by absorbing light from the stars that happen to lie behind it.

degenerate matter Matter made of leptons (usually electrons) so dense that it shows quantum-mechanical effects of pressure and temperature, as found in a **white dwarf**.

double star A body made up of two stars in orbit around each other; a **binary star** is formally synonymous but in usage refers to a closer pair.

dwarf planet (I) A **planet** in the Solar System – not one of the eight major planets, and not a **satellite** – large enough to have become spherical under its own gravity, like Vesta or Ceres, but not so large that it has cleared its neighbourhood of (other) minor planets (see **asteroid**) and **planetesimals**. (II) An **exoplanet** of the same type.

eccentric orbit An orbit that is appreciably non-circular; hence 'eccentricity', the amount by which the orbit of a body around its host is non-circular.

eclipse A celestial event that occurs, either when a body obscures another behind (as in a solar eclipse, or an eclipsing **variable star**), or when a body obscures the source of illumination of another (as in a lunar eclipse).

ecliptic The plane of the orbit of the Earth, and by extension the line that this plane makes when it intersects the **celestial sphere**.

Edgeworth–Kuiper belt
See **Kuiper belt**.

ellipse A geometrical figure having the shape of an elongated circle; the closed figure described by the orbit of a lone planet round a star.

elliptical galaxy (I) A **galaxy** of stars that has an elliptical shape on the sky. (II) A galaxy that has the form of a three-dimensional triaxial ellipsoid, which looks elliptical when projected onto the plane of the **celestial sphere**.

epicycle In orbital theory, an imaginary circular orbit, which is itself performing a circular orbit.

escape velocity The speed required for an object to escape from the gravitational pull of a planet, star or other body if it is projected straight upwards.

esker A geological feature consisting of a winding ridge of sand and gravel lying on the surface of a plain; it is caused by the deposition of sediment in the bed of a river that flows along a canal cut into a glacier, the sediment being left behind when the glacier retreats.

exoplanet A **planet** in a planetary system outside the Solar System.

extra-galactic nebula A term, no longer in current use, for a galaxy outside our own **Galaxy**.

Fermi Paradox The contradiction, formulated by Enrico Fermi, that intelligent life may be ubiquitous in our Galaxy, but has not yet been identified by the inhabitants of Earth.

fireball A **meteor** that shows brightly in the sky.

Fraunhofer lines Spectroscopic indications at various wavelengths in the spectrum of a star (especially the Sun) of the presence of various elements in the star's atmosphere.

galaxy A system of stars like our own **Galaxy**.

Galaxy The system of stars to which the Sun belongs.

gamma rays The most energetic form of radiation.

gamma-ray burst A burst of celestial **gamma rays** from an astronomical object (called a 'gamma-ray burster').

gas-giant planet A planet made mainly of gas, such as Jupiter or Saturn; formed and (in the Solar System, but not in many of the known exoplanetary systems) still orbiting at a great distance from the Sun, they have no solid surface. Compare **terrestrial planet**.

General Relativity Einstein's theory of gravitation.

geocentric theory The theory that the Earth is at the centre of the Solar System of planets; see also **Ptolemaic theory**.

geodesic The natural path taken by a light ray as it traverses a gravitational field.

geothermal spring A source of water heated by energy within the Earth.

giant molecular cloud A large cloud of gas and dust, so dense that molecules survive within it; from these molecules, stars and planetary systems can form.

glacial moraine Earth and rocks carried by a glacier and deposited off its end; the material accumulated by this method that is left behind if the glacier retreats.

globular cluster A cluster of stars, in the form of a sphere or globule, reckoned to be old.

gradualism The theory that geological formations are created incrementally. Contrast **catastrophism**; see also **uniformitarianism**.

gravitational lens(ing) A term used to characterize the effect of **gravity** on light rays, which produces distorted images.

gravitational redshift The loss of energy of radiation as it climbs out of a gravitational field; as visible light loses energy it becomes redder.

gravitational waves (gravitational radiation) The disturbances in space produced by changes in the gravitational field of an object, such as a **binary star**.

gravity (gravitation) The attractive force produced by all matter, each particle attracting every other across space and tending to change its motion; the force giving 'weight' to 'mass'; the dominant long-range force and so the force that is most important in astronomy.

greenhouse effect The warming effect produced by an atmosphere that is transparent to incoming light and opaque to outgoing **infrared radiation**, which thus becomes trapped.

Hadley cell, wave A large zone or wave in the atmosphere, of global proportions.

heliocentric theory The theory that the Sun is at the centre of the Solar System of planets; see also **Copernican theory**.

helioseismology The study of the oscillations of the Sun.

Hubble Deep Fields Observations made by the Hubble Space Telescope of typical regions of cosmic space.

hydrogen The simplest element and primary constituent of the Universe.

infrared radiation A form of radiation, slightly less energetic than light, which is emitted by warm objects.

interferometry A technique in which two separated detectors, each picking up radiation (such as radio waves, light, gravitational waves), are used to simulate (in some ways) the capacity of a telescope equal in size to the distance between them.

intergalactic cloud A cloud of gas that lies between galaxies.

interstellar dust Dust particles that lie between the stars.

interstellar hydrogen Hydrogen gas that lies between the stars.

interstellar space Space that separates the stars.

inverse square law A law that intensity or force (as in radiation or gravitational pull) diminishes according to the square of the distance.

ionosphere The spherical layer (or layers) of ionized air lying at an altitude of approximately 100 kilometres from the Earth.

irregular galaxy A **galaxy** that has no particular shape.

isothermal Of a body, having a homogeneous temperature.

isotope (from Greek: 'same' 'place' – in the Periodic Table of elements) A nucleus with a certain number of protons and a further number of neutrons. The number of protons in the nucleus of an atom defines the chemical element that it is (and therefore its place in the Periodic Table); within a collection of atoms of a given element, the number of neutrons in the nucleus of each atom varies from atom to atom. Nuclear reactions may alter isotopes from one kind to another.

isotropic Of an object, looking the same from all directions.

jet A flow of material outwards in a straight line.

jupiter A **gas-giant planet** in another planetary system that resembles Jupiter in the Solar System.

Kelvin degree A degree on the centigrade (Celsius) temperature scale, but counted from a temperature of absolute zero.

KreeP A group of elements – potassium (K), rare earth elements (ree) and phosphorus (P) – found together in some rocks in high abundance.

Kuiper belt The belt of **asteroids** and/or **comets**, or other **Trans-Neptunian Objects**, which lie beyond Neptune; hence, 'Kuiper Belt Object', one of the bodies in the Kuiper belt.

Late Heavy Bombardment (lunar cataclysm) In the early history of the Solar System, an event in which a rain of **meteorites** crashed down on the surface of the **planets**.

lava flow A flow of liquid magma from a volcano or other fissure in the surface of a planet.

lenticular galaxy A **galaxy** that is lens-shaped.

light year The distance that light travels in a year; hence also 'light minute', 'light second', etc.

log N–log S In astronomy, a mathematical expression that counts and interprets the numbers of astronomical sources of a given brightness.

lunar cataclysm See **Late Heavy Bombardment**.

lunar eclipse The event that occurs when the Moon passes into the shadow cast by the Earth in the light from the Sun.

lunar seas The magma plains that show as flat, dark areas on the Moon, once thought to be areas of liquid.

magnetosphere The region on and around a **planet** influenced by its magnetic field.

magnitude In astronomy, the brightness of a star.

mantle In relation to the Earth or a similar **planet**, the outer rocky layers wrapped around the **core**.

Medicean stars The name given by Galileo to the four satellites of Jupiter.

melt The process by which a solid, as it warms, becomes first a liquid and then a gas.

meteor The phenomenon of a streak of light and radar echoes caused by a solid body (**meteoroid**) dropping from space into the Earth's or another planet's atmosphere.

meteor shower The appearance of many **meteors** over a period of hours or days, all coming from the same part of the sky.

meteorite The surviving part of the solid body, or **meteoroid**, that causes a **meteor**.

meteoroid A solid body orbiting in space, which, when it plunges into the Earth's or another planet's atmosphere, becomes a **meteor**.

microwave radiation Radiation with a wavelength intermediate between **infrared radiation** and **radio waves**.

Milankovič (Milankovitch) cycles Oscillations in the Earth's temperature caused by variations in the planet's orbital parameters (such as its eccentric orbit or its tilt).

millimetre waves Radiation with a wavelength in the millimetre region: short-wavelength **microwave radiation**.

minor planet Obsolescent term for **asteroid**.

moon A **satellite** in orbit around a **planet** or **asteroid**.

nebula (Latin: 'cloud') A body of gaseous material and/or dust grains in space, which emits or reflects light and other energy picked up from stars nearby.

Nebular Hypothesis The central tenet of a number of theories of the origin of the Solar System, according to which the planets formed from a nebula in orbit around the newly born Sun.

neutrino A particle with no electric charge, very little mass (formerly thought to have no mass at all), and spin ½, which is produced as a by-product of numerous nuclear reactions. There are three types, or 'flavours', of neutrino.

neutron star A star so small that its constituent material is made primarily of neutrons, as opposed to electrons and protons or other nuclei; see **pulsating radio star**.

nova (Latin: 'new [star]') A star that flares up because of an explosion on its surface and becomes temporarily bright where no star was noticeable before; see **supernova**.

nuclear energy Energy released by nuclear processes; the source of star- and sun-light.

nuclear fusion A nuclear process in which nuclei are fused together to form a heavier nucleus; the process that fuels the stars.

nucleus (of an **active galaxy**) The central, small, luminous area of a galaxy, the source of much of its radiated energy, the location of its supermassive **black hole**.

oblate The term used to describe an ellipsoid that is flattened at the poles: the three-dimensional shape produced by rotating an ellipse around its minor axis; compare **prolate**.

obliquity of the ecliptic The angle (currently about 23.5 degrees) between the Earth's Equator and the plane of the Earth's orbit; thus, the tilt of the Earth's polar axis relative to its orbit.

occultation An astronomical event in which a nearby body passes in front of and hides (occults) a more distant one, particularly but not exclusively when the Moon passes in front of a planet or a star. A total solar eclipse is an occultation of the Sun by the Moon.

Olbers' Paradox The contradiction, popularized by Wilhelm Olbers, between the assumption that the Universe is infinite and eternal and the fact that it is dark at night.

Oort Cloud The hypothetical region on the periphery of the Solar System from which long-period comets come.

orbital plane The plane containing the orbit of a body that moves round another.

parabola An open geometric figure described by the orbit of a body falling into the gravitational field of another, as in the case of a comet falling in towards the Sun from a very large distance (effectively 'from infinity'). Akin to an **ellipse**, which describes the orbit of a comet that is always contained within the Solar System.

parallax The apparent shift in position of a distant object, caused by a change in the position from which it is observed. In astronomy, a shift of 1 arc second in the angular position of a star relative to its average position as the Earth moves around the Sun in one year is described as a parallax of 1 arc second, and means that the star is 1 parsec (3.26 light years) away.

perihelion The point in a **planet**'s orbit at which it is closest to the Sun.

permafrost Ground that is perpetually (or at least for an entire season) at a temperature below the freezing point of water.

phase The varying appearance of a planetary body, such as the Moon or Venus, due to its illumination by the Sun and the angle at which it is viewed; in different phases such a body will appear as a crescent, a gibbous shape, or a complete circle.

planet As defined by the International Astronomical Union, a celestial body orbiting its parent star, the body being of such a mass and size that it has become rounded and has also cleared the nearby orbital region of all the **planetesimals** among and from which it formed. Compare **dwarf planet**.

planetary nebula A **nebula** that resembles a **planet** in appearance; such nebulae were formed from material ejected by a star during its lifetime, and now illuminated by the star's core as it evolves to its end as a **white dwarf**.

planetary rings A system of **meteoroids** in orbit in a thin disc around a planet, as, for example, Saturn's rings.

planetary system A system of planets in orbit around a star, as, for example, the Solar System.

planetesimal In a newly formed planetary system, a small **planet**, formed from accreted dust and probably about to merge with others to form a large planet.

plenum A material, hypothesized by René Descartes, that fills space and transmits force (**gravity**) from one body to another.

plutino A **Trans-Neptunian Object**, whose orbit, like Pluto's, resonates with that of Neptune, and which makes two orbits to Neptune's three.

pre-biotic chemistry In organic chemistry, a process that involves complex chemicals, which, though similar to those involved in biochemistry, are not necessarily or not perhaps as complex as biological ones.

precession The gyrating motion of a planet (notably the Earth) as its rotational axis describes a cone in space; it resembles the motion of a spinning top.

principle of equivalence The principle that the mass of a body participating in gravitational attraction and the mass that resists acceleration are identical.

prolate The term used to describe an ellipsoid that is pointy at the poles: the shape produced by rotating an **ellipse** around its major axis; compare **oblate**.

proper motion The motion of a star through space across the line of sight; it is 'proper' in the sense of 'belonging to the star itself', as opposed to resulting from the way that it is viewed.

proplyd A short form for **proto-planetary disc**.

proton A fundamental particle, the nucleus of a typical **hydrogen** atom, with positive charge, spin ½ and of considerable mass.

proto-planet A newly forming **planet**.

proto-planetary disc (proplyd) A disc of dust and **proto-planets** in orbit around a new-born star.

Ptolemaic theory The **geocentric theory**, attributed to Ptolemy.

pulsar A short form for **pulsating radio star**.

pulsating radio star (pulsar) A rotating **neutron star**, showing regular and rapid pulses of **radio waves**.

quasar A short form for **quasi-stellar radio source**.

Quasi-Stellar Radio Source (Quasar; Quasi-Stellar Object, QSO) A point-like radio source coincident with a star-like optical source, indicating a very distant **active galaxy**.

radial velocity The motion of a star in the line of sight away from or towards the observer.

radiant In a **meteor shower**, the vanishing point of the tracks of the individual meteors in their orbit round the Sun, as seen in perspective from the Earth.

radiation pressure The pressure caused by light or other radiation on a solid surface.

radio galaxy A **galaxy** that emits **radio waves**.

radio waves Radiation with a wavelength upwards of about 1 metre.

red giant A large star, red because it has a low surface temperature; an evolved star.

reflection nebula A **nebula** of **interstellar dust**, which reflects the light of a very nearby or embedded star; essentially the same as a **dark**

nebula except that there is a star to illuminate it.

Roche limit The distance from a planet that constitutes the limit outside which a satellite will maintain its integrity, but within which it will break into pieces and form a **planetary ring**.

satellite A body that is in orbit around, and subordinate to another; particularly a smaller body or **moon** in orbit around a larger **planet** or **asteroid**, but also a smaller galaxy in orbit around a larger body (as with the Magellanic Clouds).

scintillation Twinkling, as of a star, due to the atmosphere, or a radio-emitting **quasar**, due to interplanetary plasma.

Seyfert galaxy An **active galaxy**, showing spectroscopic evidence of gas in an **accretion disc** and a **circum-nuclear disc**.

solar eclipse An **occultation** in which the Moon passes in front of the Sun, and may obscure it partially or totally.

solar flare An explosive event on the Sun's surface.

solar nebula The **nebula** that surrounded the Sun at its formation.

solar neutrino problem The question of why there is a scarcity, below original expectations, of solar **neutrinos**, which is answered by the theory of neutrino oscillations.

solar wind The outward flow of material from the solar surface into the rest of the Solar System.

space weather The influence of the **solar wind** on the terrestrial environment.

Special Relativity Einstein's theory of phenomena that are particularly noticeable when something is moving near the speed of light.

spectroscopy The science of analysing spectra (see **spectrum**); hence 'spectroscope' and 'spectrograph', instruments that perform such analyses.

spectrum The range of wavelengths across which an object emits radiation.

spiral arm An outward-curving component of a **spiral galaxy**.

spiral galaxy (spiral nebula) A **galaxy** of stars with a pronounced spiral pattern.

splosh crater A **crater** on Mars that is surrounded by a lobed pattern, apparently produced by the impact of a **meteorite** falling on an area of fluidized mud.

sporadic meteor A meteor that has no correlation with others seen at the same time; compare **meteor shower**.

standard candle A star or other luminous phenomenon that has the same brightness in all circumstances and everywhere; used in astronomy to determine the distance to distant star clusters or galaxies.

star cluster A cluster of stars, formed together at the same time and continuing to exist together.

static (radio noise) An emission of radio energy across a broad band of frequencies.

steady-state theory The theory that the Universe has been and always will be the same; see also **continuous creation**.

sublimate The process by which a solid, as it warms, becomes directly a gas. Contrast with **melt**.

sungrazer A **comet** that approaches very close to the Sun – in some cases so close that it entirely melts.

supermassive black hole A **black hole** formed from the amalgamation of many stars.

supernova A major stellar explosion that causes a particularly large burst of light and a much brighter than usual **nova**.

supernova of Type I A **supernova** that shows no sign of **hydrogen**; in particular, a supernova of Type Ia, which is the explosion of a **white dwarf**.

supernova of Type II A **supernova** involving **hydrogen**: the explosion of a massive star.

tail In astronomy, the material ejected by a **comet** and pushed back behind the comet's nucleus by **radiation pressure** and other forces emanating from the Sun.

terrestrial planet A **planet,** such as the Earth, made of solids and having a solid surface; compare **gas-giant planet**.

thermokarst A geological landform produced by the melting of a region of **permafrost**.

time dilation An effect of relativity (see **special relativity**) that causes time to run slowly.

Titius–Bode Law See **Bode's Law**.

torus A ring shape, like that of a doughnut with a hole in the middle.

Trans-Neptunian Object (TNO) An **asteroid** or **comet** that orbits beyond the planet Neptune; see **Kuiper belt**.

Tychonic theory The theory, articulated by Tycho Brahe, that the Sun and Moon orbit the Earth, stationary at the centre of the Solar System, and the other planets orbit the Sun.

unification theory A theory that various sorts of active galactic nuclei (see **active galaxy**) are all fundamentally the same structure seen from different angles.

uniformitarianism In geology, the theory that geological formations are created incrementally; the geological equivalent of **gradualism,** as opposed to **catastrophism**.

Universe The entirety of creation; the largest object of scientific investigation, studied through the methods of cosmology.

vacuum energy See **dark energy**.

Van Allen belts Zones of particle radiation (**cosmic rays**) that encircle the Earth, located in its **magnetosphere**.

variable star A star that changes brightness, such as a **Cepheid**, a **nova** or a **binary star** that eclipses.

weakly interacting massive particle ('WIMP') A hypothetical particle, not very interactive with others, that could contribute to the amount of **dark matter** in the Universe.

white dwarf A small, hot star that supports itself by the pressure of **degenerate matter** in its core.

X-ray binary star A **double star**, at least one member of which is an **X-ray star**.

X-ray star A star that emits **X-rays**.

X-rays Energetic radiation, a little less energetic than **gamma rays**.

zodiac The constellations that lie along the **ecliptic**.

Further Reading

The following (English-language) books have been selected for the reading list because: they extend the historical stories recounted in this book; they provide rich material at an accessible level to amplify and explain the science of the discoveries that are explained; they are on similar material with high pictorial content; or they are works that are referred to repeatedly in the text and which are accessible to a general reader, at least in part.

Books
HISTORY OF ASTRONOMY

Aratus of Soli. *Phaenomena*. Describes the classical (northern) constellations as largely used today.

Galileo Galilei. *Starry Messenger*. Galileo describes the discoveries that he made with his first telescope.

James Gleick. *Chaos: Making a New Science*. Penguin, 1988. A popular account of the history and theory of Chaos.

Thomas Hockey (ed.). *Biographical Encyclopaedia of Astronomers*. Springer, 2nd edn 2014. Contains 1,900 entries, one on nearly every historical astronomer of any significance. Available online.

Michael Hoskin (ed.). *The Cambridge Illustrated History of Astronomy*. Cambridge Univeristy Press, 1996. Authoritative, well-illustrated and comprehensive, though necessarily a brief treatment of each topic.

Michael Hoskin. *The Herschel Partnership*. Science History Publications, 2003. William viewed by Caroline.

Walter Isaacson. *Einstein*. Simon & Schuster, 2007. His life and work.

Stephen Jaki. *The Milky Way: an Elusive Road for Science*. Science History Publications, 1973. History of theories of the Milky Way.

Robert Jungk. *Brighter than a Thousand Suns*. Harvest Books, 1970s. The history of the atomic scientists, including those who discovered nuclear fusion in stars.

Henry King. *History of the Telescope*. Griffin, Sky Publishing, 1955. What the title says; comprehensive.

Constance Lubbock. *The Herschel Chronicle*. Herschel Society, 2009 (reprint of the 1933 CUP edition). The family history, by William Herschel's granddaughter.

Simon Mitton. *Conflict in the Cosmos: Fred Hoyle's life in science.* Henry Joseph Press, 2005. An illuminating scientific biography of a controversial twentieth-century figure.

Patrick Moore. *The Planet Neptune: An Historical Survey before Voyager.* Praxis, 1966. Popular stories and popular science, before the space-age discoveries.

Joseph Needham. *Science and Civilisation in China*, vol. 3, section 20, 'The Sciences of the Heavens'. Cambridge University Press, 1959. The first and most comprehensive history of astronomy in China – now showing its age but an essential resource for anyone interested in this topic.

Isaac Newton. *Principia.* Not an easy read, but the preface describes Newton's outlook on science. Andrew Motte's English translation of 1846 is freely available online.

Colin Ronan. *Galileo.* Weidenfeld & Nicolson, 1974. The man and the astronomy.

Otto Struve and Velta Zebergs. *Astronomy of the 20th Century.* Macmillan, 1962. The half century in which the astrophysics of the stars and galaxies began, authentic science by a participant.

Stephen Toulmin and June Goodfield. *The Fabric of the Heavens.* Hutchinson, 1961. Cosmology from Babylonian and Greek classical times to Newton.

Richard S. Westfall. *Never at Rest: A biography of Isaac Newton.* Cambridge University Press, 1980. The definitive biography; a big read.

GENERAL ASTRONOMY

Daniel Fischer and Hilmar Duerbeck. *Hubble Revisited.* Copernicus Springer, 1998. Pictures from the Hubble Space Telescope, and the science.

Stephen P. Maran. *Astronomy for Dummies.* Wiley, 1999. Astronomy at a basic level, clear and comprehensive.

Jacqueline Mitton. *Cambridge Illustrated Dictionary of Astronomy.* Cambridge University Press, 2007. Accessible, complete and comprehensive. If the term is not in the glossary of the present book, it is likely to be in this one.

Paul Murdin and Margaret Penston (eds). *The Firefly (Canopus) Encyclopaedia of Astronomy.* Firefly (Canopus), 2003. Condensed version of *The Encyclopaedia of Astronomy and Astrophysics*, suitable for the more general reader.

Paul Murdin (ed.). *The Encyclopaedia of Astronomy and Astrophysics.* Nature Macmillan, 2001. Comprehensive at professional level, accessible. Contains definitions, short biographies and historical articles. Available online.

COSMOLOGY

Peter Coles. *Cosmology.* Oxford University Press, 2001. A very short introduction, by a well-known cosmology teacher.

Martin Rees. *Before the Beginning.* Simon & Schuster, 1988. Why the universe is like it is, by one of the world's leading astronomers.

Martin Rees. *Just Six Numbers: The Deep Forces that Shape the Universe.* Phoenix, 2000. Other reasons why the universe is like it is.

Joseph Silk. *A Short History of the Universe.* Scientific American Library, 1997. Introduction to cosmology for the general reader by a renowned cosmologist.

Simon Singh. *The Big Bang.* Fourth Estate, 2004. Accessible history and science of cosmology by a mathematician known for his popular-science writing.

George Smoot and Keay Davidson. *Wrinkles in Time.* Little, Brown & Co., 1993. First-person account of the discovery of the fluctuations in the Cosmic Microwave Background.

STARS, NEBULAE, GALAXIES

James Kaler. *Cambridge Encyclopedia of the Stars.* Cambridge University Press, 2006.

Robert P. Kirshner. *The Extravagant Universe: Exploding Stars, Dark Energy, and the Accelerating Cosmos.* Princeton Science Library, 2004. Supernovae and cosmology by someone intimately involved in the science.

Sun Kwok. *Cosmic Butterflies.* Cambridge University Press, 2001. The colourful mysteries of planetary nebulae explained by an astrophysicist.

Brian May, Patrick Moore and Chris Lintott. *Bang!* Carlton, 2006. Modern astronomy by two astronomers and an astronomer turned rock legend (in the band Queen) turned astronomer again.

Fulvio Melia. *The Black Hole at the Center of Our Galaxy.* Princeton University Press, 2003. Our not-so-supermassive black hole.

Paul Murdin and Lesley Murdin. *Supernovae.* Cambridge University Press, 1985. The science, the history, the place in literature of supernovae.

Paul Murdin. *End in Fire.* Cambridge University Press, 1990. Supernova 1987A.

Govert Schilling. *Ripples in Spacetime: Einstein, Gravitational Waves, and the Future of Astronomy.* Belknap Press, 2017. Buckle up for a breathtaking ride through the science.

THE SUN AND THE TERRESTRIAL MAGNETOSPHERE

Robert H. Eather. *Majestic Lights.* American Geophysical Union, 1979. The aurora in science, history and the arts.

Kenneth Lang (ed.). *The Cambridge Encyclopedia of the Sun.* Cambridge University Press, 2001. Comprehensive and well-illustrated account of the science.

Mark Littmann, Fred Espenak and Ken Willcox. *Totality.* Oxford University Press, 3rd edn, 2008. Everything about solar eclipses.

PLANETS

David M. Harland. *Water and the Search for Life on Mars.* Springer Praxis, 2004. Mars exploration.

David M. Harland. *Exploring the Moon: The Apollo Expeditions.* Springer Praxis, 1999. What happened, and what science was brought back.

Kenneth Lang, *The Cambridge Guide to the Solar System.* Cambridge University Press, 2003. The Solar System in the space era.

Kristin Leutwyler. *Moons of Jupiter.* W. W. Norton, 2003. Stunning pictures, informative text.

Dana Mackenzie. *The Big Splat.* Wiley, 2003. How our moon came to be.

Paul Murdin. *Full Meridian of Glory.* Springer Copernicus, 2009. The size and shape of the Earth.

Paul Murdin. *Rock Legends.* Springer, 2016. Asteroids.

COMETS AND METEORS

Kathleen Mark. *Meteorite Craters.* University of Arizona Press, 1987. History of terrestrial craters.

O. Richard Norton. *The Cambridge Encyclopedia of Meteorites.* Cambridge University Press, 2002. Comprehensive and well-illustrated account of the science.

Roberta Olsen. *Fire and Ice.* Walker & Co., 1985. A history of comets in art.

Nancy Southgate. *A Grand Obsession: Daniel Moreau Barringer and his Crater.* Barringer Crater Co., 2002.

Donald Yeomans. *Comets.* Wiley, 1991. A chronological history of observation, science, myth and folklore.

RADIO AND X-RAY ASTRONOMY, PULSARS, BLACK HOLES

David H. Clark. *The Quest for SS433.* Viking, 1995. First-hand account of a scientific discovery in radio, optical and X-ray astronomy.

Kitty Ferguson. *Prisons of Light.* Cambridge University Press, 1996. Black holes in stars and galaxies.

Geoff McNamara. *Clocks in the Sky.* Springer Praxis, 2008. The story of pulsars.

W. T. Sullivan III. *The Early Years of Radio Astronomy.* Cambridge University Press, 1984. Reflections fifty years after Jansky's discovery by the founders of radio astronomy.

W. Tucker and Riccardo Giacconi. *The X-ray Universe.* Harvard University Press, 1983. Inside the start of X-ray astronomy.

Gerrit L. Verschuur. *The Invisible Universe Revealed.* Springer, 1987. The story of radio astronomy, by a radio astronomer.

EXOPLANETS AND EXTRATERRESTRIAL LIFE

Barrie Jones. *Life in the Solar System and Beyond.* Springer Praxis, 2004.

Michel Mayor and Pierre-Yves Frei. *New Worlds in the Cosmos: The Discovery of Exoplanets.* Cambridge University Press, 2003. Direct-from-the-scientist account of the discovery of the first exoplanets, spoilt by a rather poor translation.

Michael Perryman. *The Exoplanet Handbook.* Cambridge University Press, 2nd edn, 2018.

Elizabeth Tasker. *The Planet Factory: Exoplanets and the Search for a Second Earth.* Bloomsbury Sigma, 2017. The story.

Peter D. Ward and Donald Brownlee. *Rare Earth.* Springer Copernicus Books, 2000. Why complex life is uncommon in the universe.

Journals

A number of journals provide material of the same sort as the books above, and their back issues are worth browsing. Their websites offer a certain amount of free material, but it is a lucky dip.

Astronomy. A popular US-based astronomy journal (monthly).

Astronomy and Geophysics. The journal of the UK's Royal Astronomical Society (bimonthly). *Astronomy Now.* A popular UK-based astronomy journal (monthly).

Journal for Astronomical History and Heritage. Australian-based (3 issues per year), authoritative.

Journal for the History of Astronomy. UK-based (4 issues per year), authoritative.

Mercury. The magazine of the Astronomical Society of the Pacific (4 issues per year, digital only).

New Scientist. UK-based magazine covering the whole of science at an accessible level (weekly).

Scientific American. US-based magazine for the whole of science in which professional scientists write in well-edited articles about front-line research at an accessible level (monthly).

Sky and Telescope. US-based magazine for astronomy enthusiasts, with a high editorial standard (monthly). News, sky-watching, science, history.

Websites

The SAO/NASA Astrophysics Data System (ADS). http://adswww.harvard.edu One of the most valuable sources of astronomical research literature, very little known outside the professional community, although freely available. Not only does it contain bibliographic entries on virtually the entirety of astronomy, in many cases the full original articles are accessible. There are extensive volumes of older material, and everything is text-searchable. Most of the entries are professional research articles, but there are also obituaries, review articles and other more accessible material.

Nobel Prize
http://nobelprize.org/
The site contains extensive material including citations and autobiographies of Nobel Prize winners.

Kavli prize
http://www.kavliprize.no/
The Kavli Prize has been established to cover fields outside the Nobel Prize and includes astrophysics. The site contains citations and biographies, but the Prize is newly established and the number of recipients is relatively small.

Astronomiae Historiae
http://www.astro.uni-bonn.de/~pbrosche/astoria.html
Website for the history of astronomy, showing its age but still useful. The site includes a comprehensive list of journals that publish the history of astronomy.

Hubble Space Telescope
http://hubblesite.org/.
Highlights from the Hubble Space Telescope. There is an archive of press releases, and a gallery of pictures, including the spectacular Hubble Heritage collection, as well as sections on the telescope, astronomy and the HST's telescope, and the James Webb Space Telescope.

NASA Science programmes
https://science.nasa.gov/
A portal to all the science missions run now or in the past by NASA.

ESA Science programmes
http://sci.esa.int/home/
A portal to ESA's science missions, past or present.

European Southern Observatory
http://www.eso.org/public/outreach/pressmedia/index.html
Scientific discoveries made with ESO's telescopes.

National Radio Astronomy Observatory
http://www.nrao.edu/index.php/learn
The world of radio astronomy.

Wikipedia
http://en.wikipedia.org/wiki/Category:Astronomy
Astronomy is one of the higher-quality sections of Wikipedia, an amazing resource that can be freely edited by anyone, and is thus kept up to date by enthusiasts.

Astronomy Picture of the Day
http://antwrp.gsfc.nasa.gov/apod/astropix.html
Virtually an astronomy course through spectacular pictures, one a day since 1995.

Acknowledgments

I would like to thank Robin Rees for his involvement with our recent book projects, including an extensive dialogue and picture research for this one; it is fun to work with him, as I have done for half a lifetime. Robin and I are especially grateful to Peter Hingley for his advice and assistance in researching images for the book. I would also like to thank the team at Thames & Hudson, a demanding, talented and hardworking group of people, including Ian Jacobs, Flora Spiegel, Gareth Walker, Avni Patel, Jo Walton and Philip Collyer; it has been a rewarding education to work with them, and the book is much better because of their participation. Finally, I would like to thank the Institute of Astronomy, Cambridge University Library and Wolfson College at the University of Cambridge, and the Royal Astronomical Society for their support, especially from the libraries, while I worked on this book and other projects for the last several years.

Illustration Credits

Index

References in **bold** indicate plate numbers